D1555077

Alien Listening

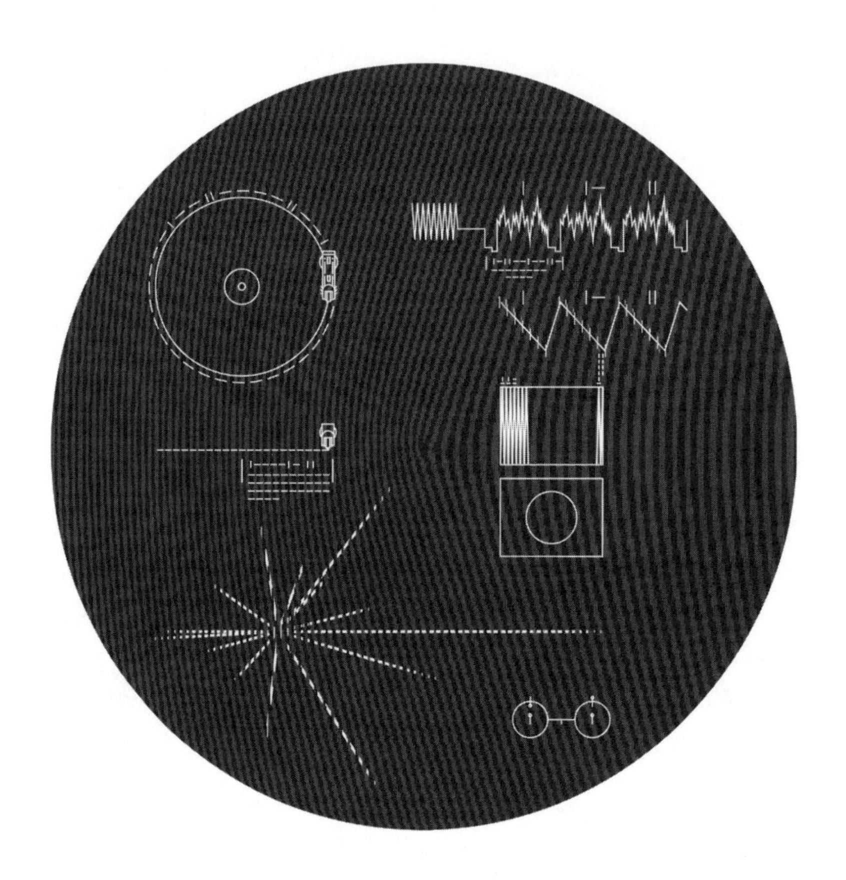

Alien Listening

Voyager's Golden Record and Music from Earth

Daniel K. L. Chua

Alexander Rehding

with illustrations by

Lau Kwong Shing and Takahiro Kurashima

ZONE BOOKS · NEW YORK

2021

ZONE BOOKS
633 Vanderbilt Street
Brooklyn, NY 11218

Printed in the United States of America.

Distributed by Princeton University Press,
Princeton, New Jersey, and Woodstock, United Kingdom

Library of Congress Cataloging-in-Publication Data
Names: Chua, Daniel K. L., 1966– author. | Rehding, Alexander, author. | Kurashima, Takahiro, 1970– illustrator. | Lau, Kwong Shing, illustrator.
Title: Alien listening : Voyager's golden record and music from Earth / Daniel K. L. Chua, Alexander Rehding ; with illustrations by Lau Kwong Shing and Takahiro Kurashima.
Description: New York : Zone Books, 2021. | Includes bibliographical references and index. | Summary: "In 1977 NASA shot a mixtape into outer space. The Golden Record aboard the Voyager spacecraft contains world music and sounds of the Earth with which humanity represents itself to any extra-terrestrial civilizations. This book asks the big questions that the Golden Record raises. Can music live up to its reputation as the universal language in communications with the unknown? How do we fit all of human culture into a time capsule that will barrel through space for tens of thousands of years?" —Provided by publisher.
Identifiers: LCCN 2020044647 (print) | LCCN 2020044648 (ebook) | ISBN 9781942130536 (hardcover) | ISBN 9781942130543 (ebook)
Subjects: LCSH: Exomusicology. | Voyager Project.
Classification: LCC ML3799.4 C58 2021 (print) | LCC ML3799.4 (ebook) | DDC 780.999—dc23
LC record available at https://lccn.loc.gov/2020044647
LC ebook record available at https://lccn.loc.gov/2020044648

Contents

Pre(inter)face

This book began on the back of a napkin in 2016. Unfortunately, this object is no longer with us, most likely because its cosmic significance transpired only many days later, after a flurry of emails between us: Could it be that our very ordinary breakfast in a corporate hotel in Vancouver was the most stimulating exchange we experienced at an annual convention of musicologists and music theorists? Sad. But maybe! For sure, it was the most extraordinary encounter. What were the chances of two scholars from distant corners of the Earth with no prior knowledge of each other's nascent thoughts on intergalactic music connecting over breakfast? It must have been cosmic providence. We regret the napkin is no longer here as proof.

Intergalactic music is no small project. Just thinking about it after breakfast ruptured our rational capacity and sent our imagination adrift in space. At first, we considered forming the Intergalactic Council of Musicologists to propagate our grandiose ideas. After all, who wouldn't want to don a cape and preside as supreme commander over an intergalactic council? Then we came down to Earth and thought an edited book would be more practical for the learned corridors of academia. The esteemed members of our intergalactic council could simply be demoted to contributors under our editorship, and that coveted cape of supreme authorship could double neatly as a book cover to contain our delusions of power. However, such a predictable genre would betray the cosmic vision on our napkin: an edited book would be too stuffy, if not a bit dusty, for a musicological space mission, and it would certainly limit our flights of fancy. So finally we decided to take responsibility for our own madness, and this jointly authored monograph is the result.

Had the napkin survived, it would have revealed a gaping hole at the core of our joint venture. This was less an accidental tear than a hypothetical placeholder. When we started the project, our thinking was devoid of any preformed theory or brand of philosophy to underwrite the mission. There was no method to frame the research (except, perhaps, the hole, which is not much of a frame). In fact, there was no research prior to the formation of the content. There was just a gap — a deliberate gap — because we decided to work backward. Rather than apply theories and methods from elsewhere to validate music's academic credentials, we began with the object itself: music. Or more precisely, music in space and its realization on the two Voyager spacecraft that NASA rocketed beyond our solar system in search of alien life. Once this music conditioned the materials for thought, its ramifications were given free rein to attract whatever theories, philosophies, and methods lay in its path until they orbited the object like the rings of Saturn.

It was an act of reverse research, simultaneously an experiment in erasure and an exercise in attention. First, we attended to the object, then we wrote about it, and finally we read, searching for the literature that would have influenced our writing had we done things the right way around. Of course, our brains were hardly blank or unbiased, but our attempt at suspension allowed the backward projection to amplify our initial materials in surprising ways. And it should be a surprise, given the extreme conditions of space. So as we worked on this book, music curved the fabric of thought in peculiar ways, attracting certain systems to spiral toward its being while leaving others adrift in the dark.

Since we reversed the normal order of things, it is vital not to mistake the curve for the object. Despite the gravitational warp in our thinking, we have no adherence to a particular school of thought, let alone any compulsion to be consistent within a system except a certain eclecticism made coherent by music. *Music comes first*. Everything else is secondary. In fact, music should transform everything in its orbit and destabilize its satellites. It might even smash and fuse them into unfamiliar compounds. Music is the center of gravity in this book.

This project, then, is about music. And the underlying question is: What is music? It is not, What makes great music? or What does music mean? or, What is the function of music in society? These are all important questions, but they get in the way of the basic question and in fact cannot be answered without first asking "What is music?" And nothing works better for decluttering the debris of definitions than to rocket music into the vast, contextless expanse of space in order to estrange music's being. A music for aliens makes for an alien music.

★ ★ ★

Breakfast in Vancouver was the first staging post on our intellectual journey. Next stop: lunch in Macau (and yet another napkin), a Friday in 2017.

On a hot and sweaty day in May, this tiny island provided an improbable place for us to consider the vast and unbounded dimensions of music in space. Yet from the perspective of intergalactic musicology, Macau's heady cultural mix seemed like the closest thing on this planet to the Mos Eisley Cantina in *Star Wars*. Since the mid-sixteenth century, this former Portuguese trading post has served as a strategic hub for Western voyagers on their way to East Asia. Which is to say, Macau is an interface—an apt site for our interstitial musings on alien contact. Media, the points of contact, became the focus of our discussion in Macau. After all, Voyager's mission would be pointless if its decontextualization of music in space did not promise an entanglement with *alter*ior contexts. Music's isolation must plug back into a bustling hubbub of mixed messages, missed calls, and strange encounters with alien forms.

Determined to avoid the glitter of Macau's casino industry, we set foot in the old part of town, which was built on the sediments of such cross-cultural encounters. We walked in the heat of the day, meandering from one historical monument to another, intermittently exchanging ideas on extraterrestrial music while admiring the eclectic mix of Qing dynasty, rococo, Cantonese, and baroque architecture. We thought of the Jesuit missionary Matteo Ricci, who arrived in Macau in 1582 to foster cultural exchange with China — in particular, his knowledge of space, in terms of both astronomy and cartography, which was much prized by the Wanli emperor and seemed prescient, given our interest in intergalactic travel. Then we needed lunch (and air conditioning). Over a meal of African chicken (galinha à Africana), the Macanese mother of all fusion dishes, which somehow melds coconut, pimenta, and peanuts in an improbable combination of tastes, we gathered our wayfaring thoughts and fleshed out the mediality of our theory.

In this transcultural context, a new set of issues emerged on our thought napkin. The underlying question was no longer, What is music? but, What is music on a planet that we will never know, for a civilization that we can never conceive, in a sensorium that may not resemble ours in any way? The ontological question became an epistemological one: less, What is it? than, How do we know? And why would it work? And this, in turn, generated a seemingly endless series of subsidiary questions that sprawled across the napkin: What happens when we can't take culture for granted? Can music still mean something when stripped of cultural mediation? Does it even need to mean anything? If not, then what does it communicate? How does mediation work under such extreme conditions? What kind of technology does an alien context demand? Do aliens even have ears? Are ears needed? And so on and so forth.

As with the methodological gap at the start of our journey, our destination is also occupied by a gaping hole. Aliens are imponderables. They represent absolute difference. Extraterrestrial contact is more a matter of miscommunication than the promise of knowledge transfer.

This book is therefore premised on an impossible mission that sets in motion a process of infinite speculation. It is as if aliens have abducted our writing, unraveling the very pages we hope to bind.

How do you write an impossible book? Indeed, by academic standards, what you are reading now may not even constitute a book. It is more an explosion of debris, a product of the tension between the gravitational force of music pulling ideas within its orbit and the speculative push of alien encounters that risks spiraling out of control.

However, if our book is premised on the impossible, then so is Voyager's mission. The impossible didn't stop NASA from turning their speculative musings into a music machine, launching their most celebrated mission after the moon landings in the process. Voyager's pragmatism provides a model for the impossible. When the space agency launched music into space, it required its alien recipients to assemble a makeshift gramophone from bits of the spacecraft in order to play back a record that, in itself, is an assemblage of disconnected music mixed with other materials in different media — images, texts, sounds. The coherence of an impossible mission comes in bits and pieces, and our book mimes NASA's pragmatism of the impossible. It, too, is an assembly of things. It is mixed-media data storage — an agglomeration of texts, equations, premises, cartoons, axioms, a manifesto, a sign-up form, rules, moiré patterns, dots, dashes, blanks, diagrams, and anagrams.

We hope this assemblage will somehow all hang together for you. Its cohesion is likely to be precarious. But what else would you expect from a book made on an assembly line? For a start, there are two authors, based in different continents, trying to slot and bolt units of thought together across time zones and land masses; then there is a comic artist from Hong Kong, Lau Kwong Shing, brought in to short-circuit the text with visual one-liners; and finally, there is Takahiro Kurashima from Japan, whose artworks are literally interference patterns that ripple across the book to illustrate our points.

Given such a modular manufacturing process, some kind of quality control was needed, and so this book was tested under extreme conditions by scholars who would have been chosen as members of our intergalactic council had it materialized under our supreme rule: Doug Barrett, Raphael Bousso, Michael Brownnutt, John Casani, Emil de Cou, Ann Druyan, Nina Eidsheim, Mark Holland, Youn Kim, Gray Kochhar-Lindgren, Lawrence Kramer, Jon Lomberg, Robert Peckham, John Durham Peters, Evander Price, Somak Raychaudhury, Jennifer Roberts, Gavin Steingo, Jonathan Sterne, Gary Tomlinson, Bert Ulrich, Margaret Weitekamp, Jennifer Wiseman, and the many other people who patiently listened to us hold forth about our intergalactic theories of everything. We are indebted to all these scholars and artists for helping us assemble this item. Needless to say, what works is theirs to boast, and what falls apart is ours to own. Although, if any bits fall off while reading, you are welcome to keep them and run off with the ideas.

We also owe a debt of gratitude to the editors of Zone Books. A special shout-out goes to the design team around Meighan Gale and Julie Fry for a graphic layout truly out of this world. Bud Bynack, our copyeditor, is an absolute star, radiating his laser-sharp light into the darkest recesses of our grammatical universe. And special thanks go to Ramona Naddaff for providing us the leeway to flex our left-field ideas in the right way and turn our tea-stained napkin into this book-like reality.

Considering the vastness of our subject, it is amazing how much little things matter in the process of writing of this book. NASA dedicated its tiny LP to the makers of all music, in all time and all worlds. We would like to invert this and dedicate our biggest ideas to the little makers of music in the cozy universes that are our respective homes in Hong Kong and Washington: Lucas and Harrison, and Emmy and Benjamin. The future of the planet is firmly in their hands. And of course, this project would have been impossible without the loving attention of our better halves, Jennifer and Bevil, who made time for us to work on this book. Finally, we need to thank the stellar team of administrative and support staff at Harvard's Department of Music, especially Chris Danforth, Eva Kim, Karen Rynne, and Nancy Shafman, as well as the transcendent staff at the Loeb Music Library around Sarah Adams, with a special shout-out to Lingwei Qiu. Iris Ng and Karen Leung in the School of Humanities office at the University of Hong Kong were invaluable in processing all kinds of paperwork to facilitate research on this project. We also wish to thank Belinda Hung, whose endowment of the Mr. and Mrs. Hung Hing-Ying Professorship enabled the commissioning of the artwork for this volume. With everyone's support, we felt that the force was with us.

Daniel K.L. Chua, *Hong Kong*

Alex Rehding, *Washington, DC*

INTRODUCTION

Blink Bang

Space, the final frontier. These are the voyages of the starship *Enterprise.*
Its continuing mission: to explore strange new worlds, to seek out
new life and new civilizations, to boldly go where no one has gone before.
—*Star Trek*

OOOI. BACK TO BASICS

Space.

Imagine a dot in space.

Space without a dot is nothing. Adding a dot transforms nothing into space. It makes space. Space emerges from the dot in relation to other objects.

And now, to make some time. Imagine a dot blinking in space — on and off — repeating its rhythm in any time.

As with the dot, the blink in time is not contained by something. It is not an event that happens in a container called "time." The blink is itself a timing that shapes time in relation to other timings. Time emerges from the blinking, just as space emerges from the dot. A dot blinking is, therefore, a *making* of time and space. Connecting the dots is the possibility of rhythm.

In this book, we imagine music as a dot blinking. Making time, making space. Weaving them together. That's it.

Perhaps this sounds too simple, too blinkered. But keeping things simple is the rhyme for the rhythm in these pages. The forbidding complexities of music theory are not necessary at all.[1] These pages are not designed to be a closed book targeted at the musically initiated and theoretically literate. Rather, in order to include everyone, this book is an attempt to retheorize music from its minimum conditions.

So, to begin, here is a working definition: music blinks.

Simple.

0010. INTO THE DARK

But not that simple.

The idea may appear minimal, even too straightforward to warrant a theory, but by "space" we do not only mean a location, like a dot on a page; we also mean "space, the final frontier," as if that dot were a pulsar or a space probe twinkling on and off as it catches the light of a distant star.

Why? Because this book is an *exo*musicology. It explores music in space. It has intergalactic, exoplanetary ambitions. It wants to communicate with aliens.

So let's reimagine this little dot in space.

Things become slightly more complicated. The conditions for music to exist may be minimal, but the context is now vast. In fact, the magnitude of space can hardly be described as a "context," since space is beyond contextualization. It is alien territory. Unknown, unknowable, a final frontier without finality. Space and time — or more properly "space-time" — can no longer be contained and controlled by human thought. It becomes mind-blowing. The blinking dot, which is so simple to grasp, travels in a space-time dimension that is inconceivable. It exists under extreme conditions on a scale that exceeds our little blip as a species on this blue planet.

Music in space would look something like this:

Did you see anything? Any little dot blinking in space? Because there is no dot there to see — not even on a nanoscale. We deliberately drew a blank.

It is not that there is nothing out there. Far from it. Maybe you imagined something hidden in the texture of the paper, an accidental inclusion or crinkle that winked at you like a dot; maybe you felt something vibrating on the page, weaving its time across the surface. Maybe... but, given a distance measured in light years, who knows what is on our symbolic black canvas? Humans are irrelevant on this vast, immeasurable scale. Anything on the page of the universe, real or imagined, is uncertain.

We drew a blank to remove the human from the picture.[2] We no longer know what is out there, because we are not there; we don't get to see the dot. It's blank. By bracketing the human, we have posited a flat, speculative "space." The final frontier has become a site for what Ian Bogost would call an "alien phenomenology." It begins by waving goodbye to our "solitary consciousness" on this planet in the same way that the "Voyager spacecraft leaves behind the heliosphere on its way beyond the boundaries of the solar system."[3] We become very small and then disappear in the vastness of space. What appears instead is a strange universe with a uniform distribution of everything. Yes, *everything* — which is quite a lot, considering the blankness of the page.

And look — there we are! We, humans, are actually there, since we are included in everything, but our thereness is simply as a thing among other things, entangled in other beings, relating as an object with other objects. But these things no longer exist for us; the universe is not one giant selfie with our face in the middle of it. The human subject is just another object on a flat surface that is aptly known as a "flat ontology," where all objects exist equally on the page without human taxonomies and privileged hierarchies. An alien phenomenology is a knowledge that goes native by becoming alien, charting how nonhuman entities experience a "flat" world. Aliens are everywhere.

So, what is music in a flat ontology?

OOII. GOING FLAT

Imagine a frog.

. . . a frog croaking on and off. Alongside the frog, add Beethoven—perhaps his "Moonlight Sonata." And more things: Elvis, Bo Ya 伯牙, Lady Gaga, a cassette tape, *kapuka*, a whale song, a performance of *Don Giovanni* tingling in Kierkegaard's ear, a laughing kookaburra, "Deus Creator Omnium," Anuradha Paudwal, a score, a vuvuzela, ABBA, presapiens flint knapping, a C-major triad, a siren, a shakuhachi, the Jackson 5, a DX7, K.271, 4′33″, the ringing frequencies of interstellar clouds, Monteverdi's *Vespers*, that annoying cell phone in Carnegie Hall, football chants, Édith Piaf, Billie Holiday, Johnny Hallyday, *Hello Dolly!*, the twang of an oud, the soundscape of Bogotá, a Swiss cowbell, the jangle of dry leaves in trees, hocketing primates, a harpsichord, a tabla, Rachmaninov's Second Symphony, prayers from a minaret, and, of course, the greatest bit of space bling in human history: NASA's Golden Record of our planet's greatest hits, mounted on the Voyager spacecraft, shown in Figure 0.1, which left us far behind as it exited the solar system many moons ago.

Figure 0.1. Voyager 1. Voyager's Golden Record is mounted on the outside of the space probe (image: NASA/JPL).

We could include a litany of other music—*even music that we cannot imagine or comprehend*—because our dark ontological surface also doubles as an image of "black noise," which, unlike the searing buzz of white noise, is barely audible. Black noise traces the quiver of things we fail to perceive. The world hums with the background hiss of "muffled objects hovering at the fringes of our attention."[4] All things, however marginal, possess a vibrancy, as Jane Bennett might put it.[5] Our seemingly vacant page, in all its eerie blackness, is brimming with a peripheral music—a vibrating poetics of everything. Everything is dotty. Everything is blinking. This music is both an object (the "thing" in everything) and a vibration (the "ing" that turns a noun—say, "blink"—into a verb: "blinking"). It is both an undisclosed item cloaked in an impenetrable darkness and energy radiating out into the unknown. If black noise is the background frequency of a flat ontology, then everything is potentially music, or at least everything is musically entwined in a pulsating mesh in which all objects are interwoven. An alien phenomenology of music roams this ontological surface to "amplify the black noise of objects" and to sound out the inner resonances of things unknown. "Reality," states Niklas Luhmann, "is what one does not perceive when one perceives it."[6] In methodological terms, to navigate this reality, musicology must avoid gravitating toward the known objects of its frequent attention. Instead, it has to *speculate*—to take a fictional, but disciplined leap across the final frontier and into the dark in order to explore an unknown world of music where everything, at first, appears to be a flat hum.

0100. BINARY DISCIPLINES

Musicology has many subdisciplines. However, once you bracket out the human, the cultural, historical, ethnographic, and anthropological approaches are largely rendered redundant for the speculative task ahead. If we draw a line in space and ask the human-oriented subdisciplines to step back from the final frontier, only two would remain as potential exomusicological astronauts for an alien phenomenology:

MUSIC THEORY MEDIA STUDIES

Neither music theory nor music media is necessarily about humans. In fact, humans are mostly unnecessary. As an analogy, theory is about the "blinking dots" and media is about the page on which they are printed. The two can speculate very well without humans blocking their way as the object of knowledge.

Theory and media make an odd couple. Theory, forever the introvert, tinkers with the internal mechanism of music, whereas media, the ever-communicative extrovert, presses all its technological buttons to connect music to anything and everything out there. Despite the contrast, the pairing is odd only in the sense that a Möbius strip has an odd twist in its rotation. Theory and media are entwined as a loop that imperceptibly turns itself inside out and outside in, as shown in Figure 0.2. Theory is the innerface of music: media is the interface.

Figure 0.2. Theory-Media Möbius strip.

The innerface peers into the autopoietic, self-producing operations of music. As an autonomous mechanism, the object of theory is operationally closed. Its integrity is singular. Indeed, its singularity is hermetically sealed, enabling music to repeat itself across different media without losing its identity. Iterability, then, is the measure of music's integrity, and the task of music theory is to keep the repetition constant.

In contrast, media can never be alone, because it takes two to mediate. The interface is the contact point that transforms music into a medium as soon it connects with another object. The medium is always between. Unlike the closed operations outlined by theory, the object of media is structurally open; it circulates from one site to another; its outward-facing surface generates a network of external relations with different objects attaching themselves to the interface or emerging as a series of extensions from its contact point.[7] Multiplicity, then, is the measure of media's intervention in the transformation of music as it ripples outward through an ever-expanding network of relations.

Thus, theory has a high fidelity to music, whereas media are highly promiscuous. This dialectic enables music's singularity to materialize across a plurality of objects in a process of constant (unchanging) transformation (change). At the same time, media amplify music's internal operations, unveiling hidden qualities with each transformative coupling for theory to distill. Theory and media are, therefore, not binary opposites: they twist in and out of each other as a feedback loop.

If music were to facilitate a close encounter of the third kind—alien contact—theory would analyze what is the same across species and media would modulate the difference.

0101. A BIG BANG THEORY

All this is probably beginning to sound a bit obtuse and abstract, if not a bit dotty and loopy. So let's get down to Earth with an example—an epic one, no less: the founding myth of Western music theory.

The curious thing about this myth is that it could also have been the founding myth of music media. The myth begins with a thing. A mundane, nonmusical thing minding its own business, doing its own thing among other things. It happens to be a very Heideggerian thing in that it has been the subject of much object-oriented thought, but nevertheless, despite its philosophical kudos, it is really no-thing for a media or music theorist to take seriously.[8]

The thing in question is a hammer.

This hammer — in fact, a scrummage of multiple hammers at work in a blacksmith's yard — was doing what hammers did when ready-to-hand in ancient Greece: banging loudly, shaping metal on anvils.[9] On this occasion, however, the banging revealed to humanity a secret hidden within the density of the object's being. The normally oblivious working of the hammer suddenly announced itself as an alien phenomenon and repurposed the tool as a musical instrument, turning the metal workshop into an ancient *Hammerklavier* sonata. Of course, this clamor can be defined as "music" only in a flat ontology, because, humanly speaking, there was no intention of making any music; there were no musical instruments, composers, or musicians — just chance objects cloaked in black noise breaking into "song."

If the nonhuman artifacts are making music, then the unknowing humans, distracted in metal making, are the tools, and not the hammers. The hammers are operating the blacksmiths, who are oblivious of the music they forge. In a flat ontology, nonhuman objects, such as rocks, paper, and scissors, interact regardless of human hands; in this sense, the hammers are making music regardless of human ears, because the environment itself is all ears; it is abundant with life, tingling with receptors, and vibrant with oscillators. Music happens in an environment of entangled relations; its "thereness" can be neither extricated nor effaced without the unraveling of the entire network; humans may form part of this network, but music's identity is not isolated by humans. *Music exists.* Even if there are no ears around to hear the falling hammers or — more realistically, the proverbial tree falling in the woods — the vibrations are not merely there, but potentially audible, because information is never lost in the universe, whether it is trapped on the surface of a black hole or in the grooves of a record. It is as if someone has left a recording device in the woods that turns timber into timbre. Since, according to the laws of physics, all information is retrievable, we must always posit a recording medium that delays perception and stores it as potential to be processed, replayed, and disseminated at an unspecified later time and place. In this sense, there is always a recorder that suspends the immediate necessity of "ears."

Music can exist very well without humans tuning in. It just so happened that one day, the alien phenomenon summoned the ear of a passing human: Pythagoras. He heard a clang of consonances and dissonances booming from the smithy as different-sized tools were struck simultaneously against anvils. Intrigued by this blast of heavy metal, Pythagoras examined the hammers as if they were components of a machine. Of the five hammers, one was tone deaf, but he realized that the weight ratios between the other four different-sized hammerheads determined the music in his ear, as shown in Figure 0.3. The ratios of 2:1 (octave), 3:2 (fifth), and 4:3 (fourth) generated consonances, whereas the ratio of 9:8 (tone) produced a dissonance. It made for beautiful mathematics: the simpler the integer relationship, the more harmonious the sound. And so from these four hammers, Pythagoras distilled a speculative music theory of everything: the nonhuman world is composed of numbers arranged according to harmonious proportions. The entire universe suddenly became music technology.

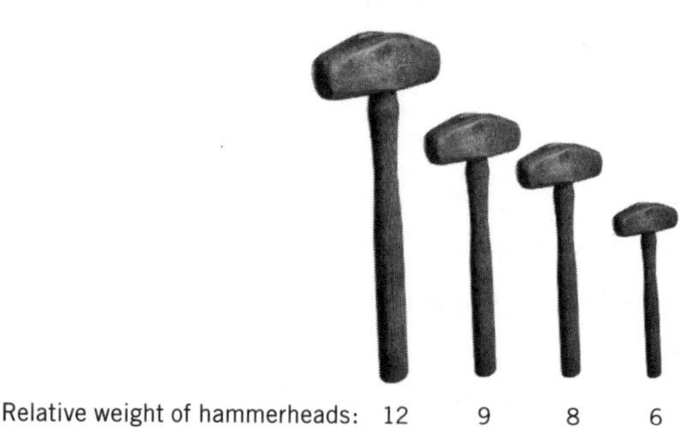

Figure 0.3. Ancient music media.

In this founding myth, the hammer, and not the human, is the agent or "actant" of music theory. By attaching itself to music's internal mechanism, the quartet of hammers became the medium for a speculative theory that transformed a universe of "black noise" into sounding numbers. Distilled as music theory, these numbers are operationally closed; they do not change, but repeat themselves across ever-changing media in a world in which the human, like the hammer, is just another "tool" — or medium — for music to operate, as seen in Figure 0.4.[10] This is because no humans made music in this myth. Rather, music makes humans into instruments, because everything is potentially its medium and everything is potentially its technology. The world is ontologically flat because music is the flat hum behind the world.

Figure 0.4. Pythagoras at work on music theory. From Franchinus Gaffurius, *Theorica musicae* (1492). According to Boethius, after the revelation at the forge, Pythagoras tested the theory: "He attached corresponding weight to strings . . . fitted other ratios to lengths of pipes . . . poured ladles of corresponding weights into glasses." In Gaffurius's illustration, the theory of numbers remains the same (16, 12, 9, 8, 6, 4) across different media (hammers, bells, glass, strings, pipes).

Strangely, given its ontological significance, music remains opaque in this system, receding into its own darkness, even as its numbers are called into the light. Despite the universal claim of theory, music always withdraws into its closed operations and conceals its being, revealing only glimpses of its inner mechanism. There is no direct contact with music, because music theory forms a loop with media and cannot escape the medium of disclosure. Music cannot be known except through its particular modification by media. So it is not Pythagoras who discovered the theory because his mind had unmediated access to music; it is the battery of hammers that banged Pythagoras's theory into shape. His idealist bent may have caused him to dematerialize music into numbers, but not before the hammers bent his ear to their "alien" being — to their shape, their weight, their volume, their velocity, their solidity, their materiality, their measure, their music.

Indeed, Pythagoras's agency in this myth *is the myth* itself, because the numbers don't add up. His calculations are mythical. The ratios are correct, but they work for strings, and certainly not for hammers. This was an immaterial point for ancient music theorists, who carried out their calculations twanging a string, rather than wielding hammers to validate the numbers. In fact, the string became the symbol of the Pythagorean cosmos. The hammers were just a founding metaphor — a big bang theory to set string theory in motion.[11] But the contingency of these found objects (the hammers) is vital to the founding myth: had these numbers originated from a string devised by human hands for the demonstration of music theory, then the universe would have lost its imperative: it would not have revealed itself by itself or inscribed its numbers in the book of nature. So although the human agent got the calculations wrong, the hammers actually got it right. The *medium is the message* in that it makes the content of theory possible — not the specific calculations, but the idea that music itself is calculation, that it has *ratio*-nality. Had the founding myth issued from the voice or from language, the medium would have framed an entirely different message — a theory of autoaffection or representation. But this would take another two thousand years or so to materialize, with the onset of Renaissance humanism.[12] The four obdurate hammers reverberating as solid metal blocks of differing sizes produced a music theory in their image in which the objective world, like the hammerheads, is ordered "according to number, measure, and weight."[13]

So with this concrete example now in hand, let's return to our abstract sketch of dots, blinks, and loops. Imagine the hammers as if they are blinking dots on the page; they form a composite machine from which space-time emerges as music, where theory and media loop in alliance with the hammerheads to explain an alien phenomenology. These banged-up, dirty, brutally ugly objects are the tiny, barely perceptible dots on a flat surface from which a beautifully designed universe came into being. Everything along its chain of being — hammers, planets, humans, and frogs — is ordered by a music that is dependent on neither sound nor human ears.

0110. NEW FRONTIERS

That was then. Pythagorean music theory, along with its well-tempered universe, no longer works. Science has changed, as has our understanding of time and space, and the metaphysics of numbers is now consigned to an enchanted past that murmurs mysteriously at the periphery of knowledge. The Pythagorean machine has become obsolete in our music history. But before we consign it to the junkyard of history, we should assess its intergalactic credentials. After all, given its object-oriented ethos and the flatness of its ontological terrain, perhaps we should salvage bits of this ancient know-how and haul it on board our musicological spaceship. It would need radical updating, of course. In a sense, our exomusicological voyage is an attempt to transform the speculative idealism of Pythagoras into the speculative realism of object-oriented thought. But this reboot would make the ancient theory unrecognizable. The new operating system would need to begin from a very different technology. (For a start, NASA rocketed an LP and stylus into space as potential found objects for alien music theory.) As for the science, this would require a fundamental overhaul, given the general malaise of musicology, which has got only as far as replacing a Pythagorean universe with a Newtonian one, where the blinking dots still exist in a container of time and space, rather than making space-time.[14] Even the calculation of dots cannot be assumed to be the same across time. Pythagoras, for example, "digitized" the harmonic ratios of 1:2:3:4 as ten dots arranged in a triangular pattern known as the tetractys—a dot matrix that underlies cosmic reality.

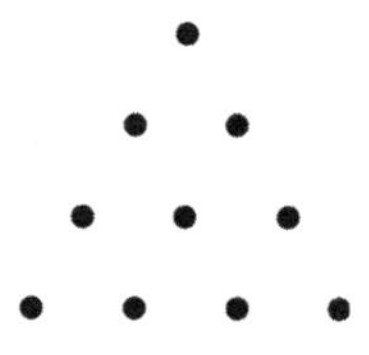

However, for the modern purposes of this book, not only will we require the tetractys to blink as a binary on-off switch, but, as in a game of billiards, we will strike the formation and cause the dots to scatter and fly to various corners of the universe as contingent and infinitely mutable sequences.

Our dots do not form a totality, but express the minimum conditions for repetition to emerge as a coherent rhythm of time. So in this reboot, there will be components of the Pythagorean machine, but there is no allegiance to a founding myth, let alone any nostalgia for an ancient universe. The task of a musicology for outer space is to open new frontiers, and not to forge old links.

The same principle of nonalliance also applies to the perspective of the new. Although we plan to retrofit Pythagorean theory with speculative realism, we have no allegiance to any one of the myriad philosophical approaches currently orbiting the object. The focus of this book is music, not a particular system of thought. Exomusicology is free to move in different directions. We merely lit a combustible mixture of object-oriented ontologies to thrust music into space, but the fuel tanks are designed to detach from the object and fall back to Earth.[15] Of course, we have our own litany of names operating as the background hum of influence—Pythagoras of Samos, Augustine of Hippo, Boethius, Hildegard of Bingen, Gottfried Wilhelm Leibniz, Johann Gottfried Herder, Søren Kierkegaard, Karl Ernst von Baer, Henri Bergson, Alfred North Whitehead, Jakob von Uexküll, Martin Heidegger, Theodor W. Adorno, Gilles Deleuze, Emmanuel Lévinas, Marshall McLuhan, Niklas Luhmann, Bruno Latour, Friedrich Kittler, Wolfgang Ernst, Catherine Pickstock, Susan Stewart, Bernhard Waldenfels, Sybille Krämer, John Durham Peters, Jane Bennett, Graham Harman, Ian Bogost, Levi Bryant, Margret Grebowicz—not to mention a laboratory of scientists who have mapped, theorized, and rocketed objects into the universe. This litany just buzzes quietly, since many of these philosophical positions are incompatible with each other and find a resonant frequency in our text only when it's necessary to delineate the object of thought. These names might surface occasionally, but having unburdened the text of the convolutions of music theory, we have no intention to tie it up in the coils and convolutions of philosophy.[16]

Keeping things simple applies as much to the methodological framework as it does to music theory, because the purpose of the book is to provide a flat ontological surface on which to map the potential of music across all disciplines in order to produce new knowledge, regardless of theoretical or philosophical aptitude. After all, if music were to reach out to life on another planet, it would not be on the basis of its philosophical or music-theoretical trappings; it would have to begin from the simplest elements. A blink, by its mere existence, is a signal. Keeping things simple, despite being very difficult for the academic mind to attain, is methodologically vital: it does not mean being reductive or naive, but being inclusive and generative to maximize the chances of making alien contact — not only with an extraterrestrial, but also with you.

In other words: we come in peace.

Simplicity may be a method, but being inclusive does not preclude being aggressive (in a well-tempered way). Yes, "we come in peace," but note: this is not a peaceful book. We have littered the text with polemics, provocations, pointed cartoons, flippant jibes, and a blunt manifesto to poke and needle musicologists from their customary, traditional positions. Our productive prodding is designed to rewire the habits of the mind, or at least to ruffle a few learned feathers. We joust and jest seriously. To "come in peace" is not about producing a text of placid prose, professionally laminated to make no difference; it is to presuppose that peace is possible in the universe, to believe that there is an ontology of peace out there, to wonder why music is possible at all — not just the music between us, but also the music between things. Otherwise, why bother with an exomusicology? Why imagine alien musicking? Why not, instead, the violence of a star-wars ontology? "We come in peace" is a provocation to reimagine the universe because such an ontology is worth fighting for. The banal cliché underlines the ethical imperative of space exploration; it is not simply a technical task. The cosmos has an ethos. NASA's prelapsarian belief in humanity is fundamentally musical, if not cosmically medieval in its nonviolent stance.[17] So, let's start by exploring the space between disciplines, in a musicology of open resonances, of hospitality, of porous difference, of synaptic connections, of alien love, of new frontiers, of constructive critique and incisive peace. This poetic incantation pulses at the heart of exomusicology. Deep down, beneath our rude textual gesticulations, we just want to save the world and make the universe a better place. Truly: "We come in peace."

OIII. GROUND PLAN FOR LIFTOFF

We also come in two pieces. To keep things simple, this book is arranged in two disciplines—or rather two faces: theory and media (innerface and interface). The general plan is to begin with the one then twist into the other. So clamped between this Introduction and the Coda are two main parts:

1. Toward an Intergalactic Music Theory of Everything
2. A Media Theory of the Third Kind

However, theory and media often merge. After all, they form a loop, with theory and media amplifying each other to define what music might be under alien conditions. You could read their oscillations as a dialectic—a tension between the general and the particular, the constant and the changing, the conceptual and the material, the innerface and interface, sound and technology—but we would prefer you to see these as blinking dots in space, the "on and off" of a binary code that winks at you in the same way that an accidental music winked at Pythagoras. Music's blink is a punctuation, a dot that punctures the page, articulating a directive as if it is looking at you. If our deployment of theory and media can galvanize an alien phenomenology of music to lure you into a close encounter of the third kind, then the mission announced in our opening epithet—"to explore strange new worlds, to seek out new life and...to boldly go where no [musicologist] has gone before"—would be accomplished.

Our two-part invention has a dual purpose. First, it attempts to open music theory from within, in the hope that stretching it to the brink of alien communication might create sufficient space for other disciplines to inhabit its universe and begin reconnecting music to everything. Then, second, it latches onto the Voyager mission in the same way that Pythagoras's attention was nailed by the hammer. After all, NASA's spacecraft carries the only music technology made by humans designed for alien listening. If we are to speculate, then our alien phenomenology might as well start from this particular thing to shape an alien media theory of music.

We have also kept the volume slim, not because we ran out of thought, but because we wanted to build a thought Time and Relative Dimension in Space (TARDIS). The book appears ordinary on the outside, but once you open its cover, you will find yourself locked in a vast interior space blinking with gadgets for infinite speculation. We are not sure where this TARDIS will take you or what kind of species counterpoint you will make with aliens, but we can promise you a surprisingly bumpy ride for a flat ontology. So as we embark on our intergalactic journey, bring along a luggage load of contingency in the hope that a few collisions with dark objects will spark some creative friction to light up your corner of the universe. Bon voyage!

TARDIS
Enter here

Instruction Sheet

In the Preface, we stated that our book is organized in "bits and pieces." Its form is an agglomeration of texts, equations, premises, cartoons, axioms, rules, dots, dashes, blanks, diagrams, anagrams, and so on.

As such, you might consider reading this book in two ways:

1. as an emergent process, where odd things connect or gravitate toward each other in unfamiliar configurations (you would need to read the book from beginning to end to experience this emergence); or
2. as a tangential process in which an item in the assemblage randomly catches your attention and sends you spiraling into another universe of ideas (in which case, you could just flip through the first few pages, then skip to the last chapter, or perhaps take a dip in the middle).

Either way is fine. There is also a third way: you can attempt both methods, either in parallel or in series. Given the variability in combining both linear and nonlinear strategies, we are providing a set of basic instructions on how you might assemble your intergalactic itinerary.

★ ★ ★

Having entered our TARDIS and arrived at this self-assembly sheet, there are now three remaining modules awaiting your engagement:

PART I
Content: Music Theory
Approach: Abstract and theoretical
Level of difficulty: 3

PART II
Content: Voyager and Media
Approach: Particular and concrete
Level of difficulty: 1

CODA
Content: Definition and Conclusions
Approach: General formulations
Level of difficulty: 2

The three modules can be configured as follows:

1. FUNNEL IN, FUNNEL OUT

We conceived the book as an emergent process, beginning with elemental particles (just dots on a page) and ending with a full-blown definition of music. In this process, Voyager slowly comes into view as you turn the corner from the theoretical abstractions of Part 1 to the concrete particulars of Part 2. If you read the book linearly from cover to cover, the emergent experience will start wide, narrow in the middle as it focuses on the concrete, and then open out into general concepts. It will look something like this:

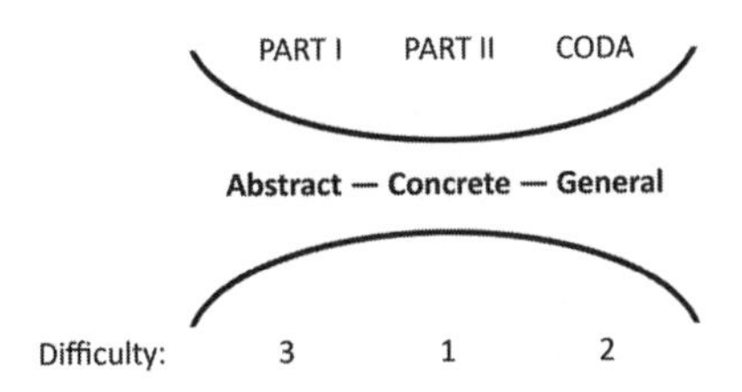

The challenge of this configuration is that you begin with the highest level of difficulty. The density and abstraction of music theory will delay your romance with Voyager.

You might object to such a difficult start. After all, we promised in the Introduction to keep music theory simple and accessible to everyone. We will keep this promise. There is nothing ahead that will require specialized knowledge of terms or symbols; there will be no jargon, no musical notation, no lengthy treatise, no exercise in harmony and counterpoint. It will be short, sharp, and simple. But simplicity should not be equated with a lack of intellectual exertion. Our theory of music is accessible to everyone, but it is not easy for everyone. It may be simple, but to be truly simple is to be truly difficult, in that such an approach demands heavy labor to ground foundational concepts. So you will have to work your way through the abstractions. It is just the nature of theory. And there is no escape: music theory is necessary, because our approach begins with music and thinks through music. We need you to become a music theorist. If you take the emergent route, this is where you start.

2. FUNNEL OUT

If the abstraction of music theory seems too much of an opening challenge, and the Golden Record is already glistening in your eye, you could reconfigure the modules so that Part 1 and Part 2 swap places. This configuration is for the reader who is under the lure of space exploration and science fiction. You begin with Voyager as a narrow focus and fan out toward the abstraction and generalities extrapolated from the concrete model. The strategy here is to pull you in at the point of least resistance so that there would be sufficient momentum to propel you through the theoretical zone and into the grand conclusions. It would be like a funnel opening outward.

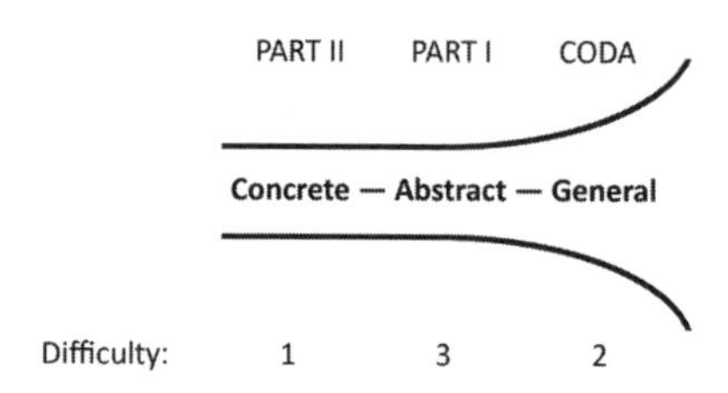

3. FUNNEL IN

Or you could turn the funnel around and begin with the end. This is for the impatient reader. Instead of an emergent book, you start by having arrived, with the outline of the destination coming first. Everything else that follows will be a flashback. Part 1 and Part 2 can come in any order, providing the details and subplots that fill out the bigger picture and fill in the gaps.

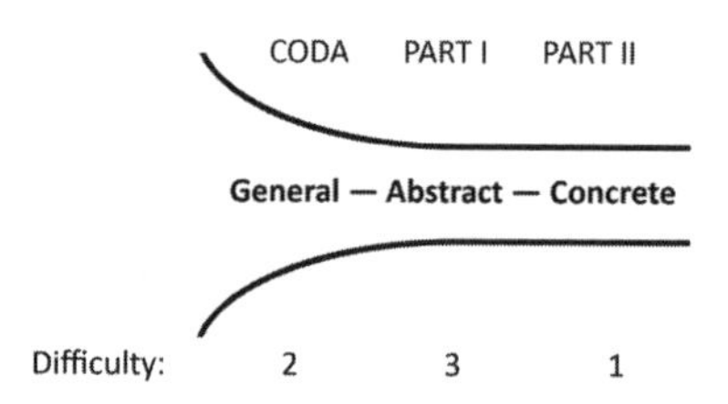

4. BIG BULGE

The Coda can also come in the middle, bloating out like a bulge of expansive ideas. In this configuration, Voyager is the concrete particular that grounds the start; it draws you into its orbit, only to open out into a grand universe of fabulous constellations. Once smitten by this dazzling brilliance, the necessary labor of music theory will seem to be no more than play. You emerge at the end as the music theorist that was always secretly your destiny.

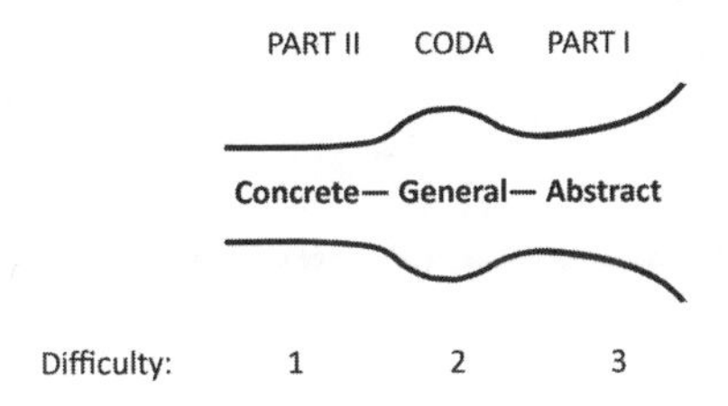

There are other possible configurations and strategies, but perhaps what is most evident in this set of instructions is our audacity in expecting you to read this book more than once. Repeat and replay is the performance we hope to elicit for you. It is implicit in the modular structure of the book: the very fact that you can reconfigure the order indicates that there are certain themes that recur in each module. Repetition is part of the experience, with each repeat viewed from a different perspective — from the viewpoint of theory, media, and their synthesis. Repetition is our attempt to turn the text into an earworm. There is no wriggling out of the process.

Torturing earworms
You do realize that if you break me,
there will just be more of me to deal with.

PART ONE

Toward an Intergalactic Music Theory of Everything

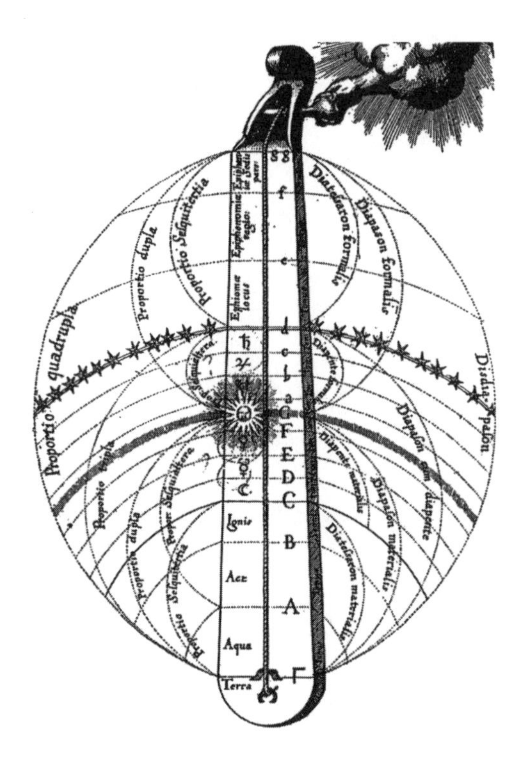

Figure 1.1. The hand of God tunes the string of the cosmic monochord that stretches from heaven to the earth to embrace all the elements within the unity of its harmonic ratios. From Robert Fludd, *Utriusque Cosmi Maioris Scilicet et Minoris Metaphysica, Physica atque Technica Historia* (1617–19).

CHAPTER ONE

Manifesto

In a galaxy far, far away...
—*Star Wars*

Music theory has seldom been modest. A manifesto entitled "Toward an Intergalactic Music Theory of Everything" should be no surprise, except for the modesty of the word "toward." Such caution is unbecoming of the genre. After all, music theory, in all its speculative glory, was the first "string theory" of the universe (Figure 1.1). Admittedly, compared with the quantum vibrations of current string theory, it was somewhat reductive, consisting of just one string; but it was a very long one that tuned the motion of the planets and ordered the entire chain of being along its harmonic series. Music theory *ratio*-nized the cosmos. It was a theory of everything. The universe vibrated with knowledge, and there was one string to rule them all.

Pythagoras would have called this his "big twang theory," were it not for the fact that such music didn't have a beginning. It was eternal, a resounding ring of timeless integers that intimated a metaphysical reality. Music was being. This ontological string pulled the world together and kept it in proportion for some two thousand years. Even when its cosmic order waned as the light of reason dawned upon the eighteenth century, music theory continued its immodest claims, stirring up quarrels among the intelligentsia concerning the identity of body and soul, matter and spirit, the world and the self. Indeed, music theory was seldom just about music. So much else seemed to hang on this one string.

★ ★ ★

Times have changed. Although music theory today is still entangled in the frayed fibers of its ancient string, it has become increasingly irrelevant in explaining anything other than itself.

A string walks into a bar

It has evolved into kind of a truism that borders on tautology: music theory is for music theorists. It has fallen into an academic narcissism that would be quite beautiful, were it not so boring for everyone else peering into the discipline. Of course, this assumes in the first place that it is possible for those on the outside to penetrate the density of its discourse in order to experience the beauty of its boredom. It is dull and forbiddingly opaque, and most scholars leave music theory alone to talk to itself.

So why bother with music theory? Because its cosmic potential is too big to fail. For music theorists, this manifesto will clearly be a sharp, short critique intended as much to be a goad as it is a call to action. For others, this internal critique may seem irrelevant. Why not skip the chapter and leave music theorists to circle within their solipsistic enclosures? Because such an omission would be a failure — a failure not just of nerve, but of method. This chapter is vital in revealing that we *all* need to be music theorists and that there is nothing to fear from the music-theoretical barriers erected to exclude a wider participation.

So what are these barriers?

To summarize the issue, there are two questions facing music theory that those on the inside often fail to see, but that are blindingly obvious when viewed from the outside:

1. Why is music theory so boring?
2. Why is music theory incomprehensible?

The two questions are obviously related, complementing each other as partial answers, but there is more to this than just boredom induced by incomprehension. Let's address the two questions.

1. In answer to the first question — Why is music theory so boring? — we need to be clear that this has nothing to do with the quality of current scholarship. It has never been better. If anything, it is too interesting to suffer from *ennui*.[1] No, this boredom is of another order. It has to do with vision. Or rather, the total lack of vision. It is as if music theory has erected thick double-dotted bar lines on all sides to contain itself in a perpetual state of internal motion (Figure 1.2).

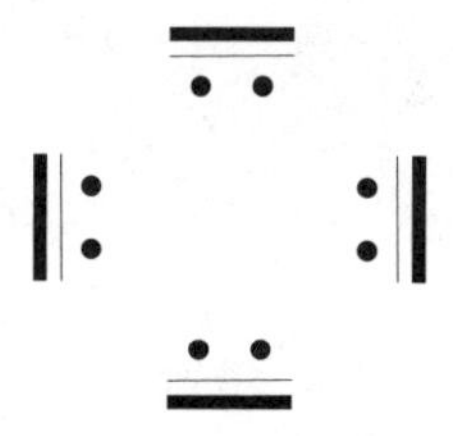

Figure 1.2. Music theory's cordon sanitaire

This isolationism probably stems from its emulation of "absolute music," the idea that music is essentially a self-referential system about only itself, music that, as Richard Taruskin explains, requires the cordon sanitaire of theory to keep music's essence pure and its form autonomous.[2] But this obsession with hygiene results in a theory that is highly allergic to everything. What was formerly a theory of everything now sneezes at the cosmos it once explained to keep its knowledge to itself. As a consequence, music theory is boring because it is irrelevant. However interesting it might be internally, it just has nothing to say beyond its little plastic bubble, its cordon sanitaire. What used to be the most gregarious and speculative of disciplines, with a reputation for nosing around the cosmos as a cross-disciplinary space invader, is now reduced to a lonely, solipsistic existence. The fact that some intrepid music theorists may burst out of the bubble, venturing beyond their disciplinary borders to enliven their own scholarship, is beside the point. The questions is: Who adopts music theory to enrich their discipline? The answer is: no one. Music theory has written itself out of any participation in epistemology. It is structurally boring.

Q: Why did the music theorist cross the road?
A: He didn't.

2. As for the second question — Why is music theory incomprehensible? — the answer cannot be attributed to the incomprehensibility of scholarly discourse, which is the native tongue of academia. Academics understand each other's incomprehensibility perfectly. No, music theory's incomprehensibility is of another order. It is usually blamed on the nature of music and the apparent opacity of its inner workings, which require the use of specialized terms, nonverbal symbols, and insular concepts that exclude most academics from listening in. But music did not invent this language. Music is everywhere and for everyone, so why is its theory so impenetrable that only a few understand it?

The problem with music theory is that it is *fundamentally* incomprehensible. What is basic is not basic — it is already too difficult. There is no passing note long enough or consonant skip high enough for the uninitiated to cross the divide and scale the walls of the citadel. To be clear, there is nothing wrong with these "basics." Who would want to hold anything against a passing note or a triad? They work just fine. After all, that's what they are supposed to do — *work*. But that is the problem. Everything *functions* in music theory: harmonic functions, formal functions, tonal functions, chord functions, thematic functions. Theory is preoccupied with work, busying itself with what music *does*, and not what music *is*. If an alien were to land inside its walls, ze (the gender-neutral pronoun seems appropriate here) would see a strange world populated with human doings, rather than human beings. The graffiti in its streets would read: "Utility rules. Ontology sucks." Its ivory towers would espouse the maxim: "To do is to be." Technique would be its law and labor its politics. But there would be no meaning, because the basic questions — the "What?" and "Why?" of being — are lost in the hum of activity.

It didn't used to be this way. In the past, music theory was all about being. That long piece of string fixed permanently at either end of the universe may not have done much, but it was the basis for everything. Everyone had their being in its vibrations. It was as simple as 1:2:3. But now, with its technical turn, music theory has become a highly specialized skill, as arcane as $\hat{3}$-$\hat{2}$-$\hat{1}$. (An in-joke for music theorists; if you didn't get it, then it proves the point.[3]) As long as theory obsesses over technique (doing) with a disregard for ontology (being), then what it considers "basic" is simple only to those already in the craft and is incomprehensible for those outside its industry. By failing to address what music is, what passes as a fundamental concept is not fundamental at all. Even something as simple as a passing note involves a highly complex operation requiring a vast technical apparatus to support its tiny steps. These "basics" are so high on the theoretical ladder that one wonders whether its base is still grounded in anything, since without "being," there can be no ground. And if you look carefully, theory has no legs to stand on. It simply uses *technique* to pull itself up by its own bootstraps. It manufactures its own standardized models, mistaking "normativity" as a form of self-rule when there is nothing "normal" supporting its claims. Hence, theory can guarantee music's autonomy only by binding its procedures with technical rules and formal laws to create a dense, disembedded, impenetrable structure. It is incomprehensible.

But this incomprehensibility points to a deeper incomprehensibility within theory itself. *Music theory does not know what music is*. Or rather, as a theory committed to technique, its "basics" are so specialized that it narrows down what counts as music; its tools are designed for a limited span of mostly Western art music and only when they are encased in the form of musical works. Theory, then, legitimizes only that which it can analyze (the canon of Western art music) and excludes most music in the world which it cannot recognize. Music is everywhere and for everyone except in music theory where it is incomprehensible—even to itself.

The Ministry of Works
Mr. Stravinsky, I believe your bi-tonal chord contravenes Section 7-32 of the ordinance for sacrificial dancing in a public space.

It is a serious situation when theory is both epistemologically insignificant and ontologically ignorant. What should be the theoretical life of music, unifying its diverse manifestations, has lost its purpose. Structurally boring and fundamentally incomprehensible, music theory has failed music. It is unable to provide a common platform from which to theorize music across the disciplines with basic concepts that are equally meaningful to all music. Music theory fails as theory.[4]

Exactly at 8:56 a.m. on September 5, 1977, the Voyager 1 space probe rocketed into space. This official launch followed on the heels of the craft's identical twin, Voyager 2, which was sent ahead a few days earlier as a trial balloon. While music theory was boldly orbiting "the music itself," the astronomer Carl Sagan sent music in the opposite direction—into outer space. A gold-plated audiovisual disc with a selection of music ranging from J. S. Bach to Chuck Berry and from Mexican mariachi music to Indian raga, as well as the wails of babies, the song of whales, brainwaves, and ocean waves was placed on board Voyager, along with a ceramic stylus-and-cartridge unit and nonverbal instructions (Figure 1.3), on how to reproduce the sounds. If nothing else, any intelligent life-form that intercepts the space probe will learn from humans how to make a gramophone before eating the disc for breakfast.[5]

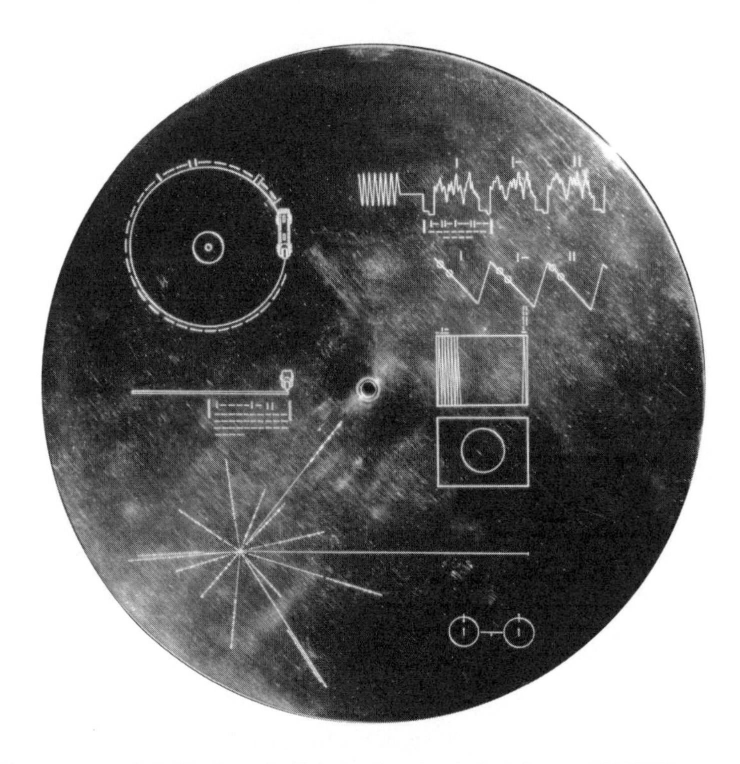

Figure 1.3. NASA's Golden Record with instructions for playback (image: NASA/JPL).

Currently, the two Voyager spacecraft are silently gliding in interstellar space, "destined," as NASA puts it, "to wander the Milky Way," although Voyager 1 is vaguely directed to pass within 1.6 light years of the star Gliese 445, and Voyager 2 within 1.7 light years of the star Ross 248. In about forty thousand years, NASA tells us, the Voyagers will be closer to those neighboring stars than to our sun.

Music theorists did not pay much attention to NASA's exploits at the time; they were probably too obsessed with "human doings" to imagine an interstellar music theory for aliens. Conversely, Carl Sagan did not include any music theory with the instructions; there was no Schenkerian fundamental structure for Beethoven's Fifth Symphony or Fortean pitch-class-set theory for the *Rite of Spring*, which would have been de rigueur in 1977 among theoretical circles. Sagan's team probably realized that this type of music theory would have been incomprehensible, if not utterly boring, to the average alien. Perhaps they also figured out that there was no single theory capable of explaining the representative samples of music encoded on the disc. If Beethoven and Stravinsky already require different theoretical models, what would a didgeridoo from Australia or panpipes from the Solomon Islands or Louis Armstrong and His Hot Seven demand of music theory? Not only is theory inadequate to the task, it would become increasingly disparate if it attempted to explain the diverse music on the record, resulting in a proliferation of discrete techniques. A golden textbook in several volumes would have to accompany the golden disc. Music theory would ultimately alienate the alien.

Earth was never music theory's home, given its cosmic pretensions in its formative years. It can never return nostalgically to its premodern existence, of course, but theory can look back to the future and reimagine a different space, knowing that its current position is not its native home. Sagan and his team expressly hoped that extraterrestrials might analyze the music of the Golden Record. So what would a music theory for aliens be like? How would another life-form begin to understand the music in that probe with the golden disc — not just the Beethoven and Stravinsky, but Senegalese percussion, rock 'n' roll, gamelan, Navajo chant, and, arguably, the music of nature itself? What would the fundamentals of this music be, given a distant galaxy with life-forms that have evolved ears in a different planetary system? Such questions, which an intergalactic music theory of everything (IMTE) poses, are clearly bizarre and appear impractical. Why posit an extraterrestrial music theory, particularly as we are unlikely to be discussing Chuck Berry with an alien at anytime soon?[6] There are two reasons.

The first reason is strategic. IMTE is needed to stop music theory from muttering to itself in a corner. An intergalactic impulse should propel theory at warp speed to the cutting edge of the humanities. Recently, the humanities have called into question the very humanity from which it derives its name. The human subject that claims to be the center of the universe is far too arrogant a being to entertain in an age where its powers have wreaked such havoc on the environment that it has inaugurated its own geological timeframe—the Anthropocene. The "posthuman" turn in the humanities is, in fact, far more human than its previous incarnation, dethroning that godlike subject and replacing it with a human who is more creaturely and more environmentally friendly—a reduced being, recyclable in time, reusable in nature, at one with biodegradable matter.

An IMTE acknowledges the posthuman and the Anthropocene, but it also whizzes pass them in its spacecraft, waving out of the window as if to say "Been there! Done that!" With IMTE, the posthuman is surpassed by the extraterrestrial, and the geological is outshone by the intergalactic; its frame of reference is measured not by thousands of earth years, but by thousands of light years. By the time some alien beams back the message "Send more Chuck Berry," the Anthropocene may be over. As a vision that extends beyond the posthuman and the Anthropocene, IMTE can have an epistemological impact from a perspective and scale that encompasses the sciences and humanities and so move beyond its own disciplinary boundaries.

If the first reason is strategic, repositioning the epistemological relevance of music theory, the second reason is practical, addressing the question of incomprehensibility. If we can design a theory that can explain music to an alien, it should be comprehensible for humans. The alien hypothesis provides a defamiliarizing frame that enables us to rethink theory from the basics. This would be a theory that has to work at any point in our universe, based on properties that we might share with an alien. It forces us to return to the fundamentals of being, of physics, of time, space, and matter. It obliges us to reevaluate what music is, particularly because any sound reproduction from Voyager's golden LP is unlikely to sound like anything we know on Earth. The differences in pressure, density, atmosphere, and evolutionary adaptation alone are enough to ensure that the Second Brandenburg Concerto — the first track on the disc — is Bach, but not as we know it. If music theory is wide enough to encompass such redefinitions of what music is, then it might finally open up a transdisciplinary platform where music can be a shared discourse that is everywhere and for everyone — on Earth.

cut along the dotted line

IMTE COMMITMENT FORM

Never thought of yourself as a music theorist? Think again. Music theory is not simply an academic discipline; it is a responsibility to care for the materials of music and makers of music in all their diversity. As such, it is too important to leave theory to talk to itself. In its current state—with its Western biases, limited range, and ingrown techniques—music theory needs to take a cosmic turn in order to recognize other cultures, listen to indigenous theories, imagine alternate worlds, and practice a hospitality that opens its discourse to others. This is a huge undertaking. It demands a revolution. Although this manifesto is aimed polemically at the institution of music theory to galvanize its potential, the task ahead is not only internal to the discipline. Music theory is not only for music theorists: it can be everywhere and for everyone. We can all be music theorists. An intergalactic music theory of everything is a democratic movement—a revolution for all. So sign up, join in, and theorize.

Intergalactic music theorists of Earth, unite!

____
Name

____
Species

____
Signature

Please check a box:

☐ I am a committed exomusic theorist. Beam me up for the mission.

☐ I would like to be an exomusic theorist. Please send more information.

☐ Unsubscribe me from this ridiculous concept.

Send to: Secretary General, The Intergalactic Music Council,
PO Box 0010, Gamma Quadrant, Sector 4

CHAPTER TWO

Blueprint

Everything is rhythm.
— Friedrich Hölderlin

OOOI. MODULES

The first problem with formulating an intergalactic music theory of everything is that no one can write it. It would be a performative contradiction. In a posthuman universe, there is no human equivalent to a music theorist such as Jean-Philippe Rameau or Hugo Riemann. No great mind can conjure up this theory, for such a definitive act would reinstall the heroic subject at the center of knowledge. Such male posturing cannot secure the autonomy of the system. Given the intergalactic scale and all-inclusive reach of the theory, no one is wise enough to know it all. To authorize, control, and canonize this theory is to fail by falling back into the arrogance of a self-centered humanism that wants to master the universe and put its name on everything. Thus, the greater the immodesty of this theory, the more modest its practice must be. Humility, not hubris, is the ethics of its pretensions "to boldly go where no theory has gone before."

All that can be produced at this formative stage is a modest blueprint to accompany the immodesty of the manifesto. This would merely be an intuitive sketch — more generative than definitive in its intentions. Some*one*, of course, would need to initiate the plan, but to build the theory would demand the work of many scholars that should render the theory more or less anonymous. An intergalactic music theory of everything is therefore not a singular theory, but a *modular* one, where different disciplines interlock to test, correct, modify, and generate the ideas. The blueprint provides an open platform of basic, editable content designed to incubate concepts that can split off in multiple directions and methods. But however complex or specialized the modules become in this process of proliferation, they should always connect with each other through the shared network instigated by the blueprint.

Summary IMTE is a modular, open-ended design for scholars in any discipline to connect and contribute to its theory.

OOIO. PARAMETERS

How does this blueprint work? As a modular theory, the IMTE blueprint begins by sketching a few basic parameters to steer the "modulations." They can be divided into rules and concepts.

Rules There are three simple rules, the first being that rules should be simple.

1. *The Rule of Simplicity* Rules should be simple. Basics need to be basic, accessible to all disciplines and equally applicable to all music. Simplicity should not be regarded as a reductive procedure or disregarded as an elementary practice: it is an inclusive and generative process. The simplest things are the hardest to theorize.
2. *The Rule of Inclusion* The theory must be inclusive. Since what is basic is foundational, there should be nothing "new" in IMTE that replaces existing music-theoretical paradigms; it must support, transform, or illuminate them. If current music theories, in their obsession with technique, nest on the higher rungs of the ladder, then IMTE at its base would fail if it simply erased their legitimacy by knocking them off their perch; it would be exclusive, rather than inclusive. A theory of everything must include everything.

3. *The Rule of Embeddedness* The basic concepts of the theory must be embedded in the universe. To be intergalactic, music must be part of the fabric of the created order; it must participate in the coherence that enables us to call the universe a "*uni*-verse" rather than chaotic gloop. Time, space, matter, life — these elements form the properties of music in which the human contribution is part of its embeddedness in nature, and not its moment of transcendence. Without such participation, music theory would have no relevance to anything on Earth, let alone to an alien life-form.

Concepts There are three concepts that parallel the three rules. These rules regulate the concepts that form the basic elements of the blueprint. The concepts therefore conform to the principles of simplicity, inclusion, and embeddedness. Together, they form a tripartite framework for the definition of music.

1. *The Concept of Simplicity* Repetition is the fundamental unit of musical coherence. Coherence is simple.
2. *The Concept of Inclusion* Repetition is measured as frequency. Music exists along a single continuous frequency spectrum. In IMTE, frequency should not be exclusively equated with pitch (as the term is often used); it covers anything that repeats in music, which is basically everything. The frequency spectrum is therefore a graded "glissando" of repetition. Elements that appear discrete or disparate are divisions along the frequency spectrum. Rhythm, timbre, pitch, and form are all related as frequency. Difference is inclusive.
3. *The Concept of Embeddedness* Time began with a bang. The universe is unfolding as a fabric of space-time. It is extending, expanding, vibrating, and folded in multiple temporalities. Music, as repetitive motion along the frequency spectrum, is a weaving of space-time. Its temporal threads loop and twist in and out of each other to form a vibrating, extendable, tensile mesh. As such,

it shares the same texture as the universe and is integrated in the weft and warp of its fabric. Music is inextricably embedded in the universe.

Summary In IMTE, the rules of simplicity, inclusion, and embeddedness are conceived musically as a theory of repetition. In this theory, music is best described in terms of loops, oscillations, turns, spirals, rotations, recursions, frequencies, and waves. Framed in this way, music functions as a universal machine or universal medium—a kind of computer through which any discipline can plug in and any intelligent life-form can engage.

The following sections of the blueprint outline, in the briefest terms, the "what," the "how," and the "why" of repetition:

- What is the nature of repetition (section 0011)?
- How does repetition work (section 0100)?
- Why is repetition important (sections 0101 and 0110)?

The purpose here is not to define a theory and fix its meaning, but to offer a few speculative coordinates to provoke further exploration on this voyage into space.

0011. REPETITION

Everything repeats.

Everything repeats. Music repeats itself endlessly. If it were a language, it would be meaningless; its interminable reiterations would be denounced as incoherent stuttering.[1] Music is not language precisely for this reason. If anything, music (en)trains language to jump through its hoops, turning its meaning into stuttering nonsense that makes perfect sense as rhythmic phenomena. For example, take a simple sentence from a song by the Police:

I can't I can't I can't stand losing
I can't I can't I can't stand losing
I can't I can't I can't stand losing
I can't stand losing you
I can't stand losing you
I can't stand losing you
—"Can't Stand Losing You," The Police, 1987

Music's repetitive motion is the basis of its coherence. Indeed, at its basic level, music is just repetition—a rhythmic fold—that holds time together as a discrete loop (see section 0100). Without this loop, music is incoherent, if not impossible. Thus, repetition is the minimal condition for music and the maximal potential for its generation. Music is therefore very simple. It is simply a matter of repetition.

Music's fundamental simplicity enables it to be a theory of everything, because everything repeats. The universe repeats itself endlessly. It operates by repetition. Through its vibrations, oscillations, waves, and rotations, it moves and measures time and space in all dimensions—from the looping membranes of string theory to the massive shudder of gravitational waves. Repetition functions as a universal in the universe—a kind of background hum that is a fundamental condition for existence. As long as there is time and space, there is repetition. Or, to put it in the terms of Fourier analysis, space-time is frequency; any*thing* that takes *place* in *time* can be expressed as frequency.[2]

This is the case for both the animate and the inanimate world. The laws of quantum mechanics and general relativity may repel each other, scuppering the scientific quest for "a theory of everything," but they at least share the same vibe: both theories require their worlds to oscillate. Indeed, in a curious validation of Pythagoras, the quantum leaps in string theory jump back and forth in accordance to harmonic ratios of a string. But to be a thing is not simply to vibrate randomly here and there, but to repeat itself in time to hold its thingness together. As Catherine Pickstock claims, to be any*thing* at all is to sustain identity as repetition.[3] As for the animate world, life also oscillates; its biochemical and cognitive mechanisms circulate, replicate, and reproduce, from neural oscillations firing across the body to the feedback loops and mimetic actions that create a social oscillation between individuals. To live is to cognize and recognize. "Life is repetition."[4]

If music is defined in terms of repetition, then it can be found anywhere in our universe, not so much because it exists in the universe, but because it partakes in the fundamental parameters of existence. Music is not contained in time, but is enmeshed in it. It is not so much a product of life as an expression of its process. Where there is frequency, there is music. Thus, any theory predicated on intergalactic communication would need to be musical, because music is woven in the fabric of life and the very dimensions of being. So although Pythagoras was wrong, he was wrong in the right way. The universe is a kind of music. You can tune in to its frequencies. To communicate across galaxies to an alien intelligence is therefore possible because we *frequent* the same space-time and life-form.

Summary

1. IMTE is premised on frequency as a shaping of space-time. Music, as repetitive motion, does not merely move in time and resonate in space, but is materially embedded in these dimensions and can therefore model and disclose their properties. Music is an aesthetics of space-time.
2. IMTE is premised on frequency as the biological and cognitive rhythm of life. Music, as repetitive motion, is embedded in mechanisms of sense and sensation, perception and reflection, motion and emotion.

These two statements are a reiteration of an ancient idea found in Greek, Indian, and Chinese cosmologies: music weaves the world together, both within us and outside us. Herder echoes this vision when he writes: "Everything, therefore, that resounds in nature is music." "It is not *we* who count and measure, but rather nature; the clavichord plays and counts within us."[5] Music embeds us in the universe, and the music we make enables us to hear how we inhabit the fabric of space and the cycles of life. To put the matter the other way around, we do not make music as its creator, but respond to a music that is already there. Repetition is therefore both an ontological domain and epistemological object of music theory.

If music weaves time both within us and outside us, then its repetitive motion operates as a mediator. Music converts time as an unknowable object into a *quality* of time that can be experienced. This relation is a metaphor: time *is* music. To listen to music is to attend to time. It is as if music scales time's immeasurable vastness into an ear-size gravitational field that warps within our being. Or, heard from the opposite end, it is as if music amplifies the subatomic resonance of the universe to dance before our ears. Thus, as a metaphor, music measures the immeasurable to make time appear as if it is calibrated to tick precisely with our internal clocks.[6] Music's repetitive tick enables us to keep time with the universe. And in turn, by making music, we are manufacturing teeny-tiny big bangs — miniature explosions of time that expand as vibrational cycles — in order to share our peculiar measure of the universe with another. And the other (perhaps, an alien), in its turn, will receive our measure of time in accordance to its own measure and perpetuate a chain of difference as music passes from one interface to another.

Repetition is a generality in the universe, a specificity in music, and a multiplicity through media.

0100. THEORY

Of course, if everything repeats in the universe, then repetition is not the sole property of music. Its embeddedness seemingly erases music of its distinguishing characteristics. Color, for example, is also frequency. Given the right ears, you could hear color. Conversely, given the right eyes, you could see music. Both can be expressed by Fourier analysis. So why single out music as a theory of everything?

What distinguishes music from everything else is that repetition is its perceived mode of communication. It makes repetition in time the focus of meaning. Thus, music not only is frequency (which makes it generic), but is *about* frequency (which makes it particular). It makes the universe audible as rhythm.

Music exists as long as its repetition is perceivable as rhythm. This does not mean that music has to be perceived in order for it to exist. Music may remain unknown. Things may not connect. However, music is *premised* on an audition that may or may not be realized. Contact is a fragile and uncertain affair. This is because music's universal "voice" can be heard only by an other who is always particular and contingent. This could be "unintentional" music (a Bolivian villager attending to a babbling brook to collect songs, for example)[7] or voluntary (such as a Bolivian mother babbling a lullaby to her newborn), but the hearing is always particularized as a transformation of the music, rather than a transmission, bringing the other in relation through difference.

Music's universal claim, then, is only in its relation—a resonance that gives access for another to enter ectopically "in time" with its rhythm. It wants to believe there is a significant other out there—whether there is anyone out there or not. NASA's mission is the ultimate expression of this desire for a close encounter: Voyager ventures into space to make our frequencies mean something long after our extinction to a being who is not only unknowable, but probably undiscoverable. This hopelessly optimistic gamble betrays our species' compulsion to repeat ourselves to anyone who cares to listen to our existence. The Golden Record, in repeating our time, literally *gives time* to another. The invitation to resonate with us, to sound us out, to syncopate with our time, expresses a proximity toward alterity that is the closest music can claim to being a transcendent language. Music reaches outward. It is always *ek*-static.

Thus, everything can be music, but not everything is music, because music presupposes the possibility of listening. Music repeats itself to be heard; it blinks to give itself away. These reiterative patterns are "mind-traps," designed to captivate the unwary passerby.[8] Thus, what makes frequency music depends on how its repetition is disclosed or received. Music's universal nature is very particular in its communication.

But why should any being attend to music's repetition? What's the catch in this musical "mind-trap" of endless reiteration? A theory of repetition is required to probe these questions.

A BRIEF THEORY OF REPETITION IN SEVEN PREMISES

1. Keeping the Piece Repetition is the fundamental component of an intergalactic music theory. As the lowest common denominator of everything, it is also seemingly the most banal basis for everything. But this is because repetition is often perceived as so "simple" that it hardly deserves to be theorized seriously, which is a mistake, since the simplest things are often the hardest to explain. If repetition is theorized at all, it is often perceived as a problem requiring justification due to humanity's obsession with language as its unique mode of meaning, a mode in which excessive repetition is seen as redundant. But far from being redundant, musical repetition is the fundamental process of making time coherent. Through musical repetition, temporality is perceived as "a piece," as something discrete within the space-time continuum we inhabit. Repetition does not require some higher structure or outside order to clamp the elements in place.[9] "Form" is not necessary. Repetition is already a "piece."

So the first theoretical premise of IMTE is that musical repetition makes "a piece" of time coherent.

2. Simple Time But why should repetition be coherent? A repeated unit could simply divide itself from its origin and fall away as some disconnected event. Just because they are the same does not mean that they are coherent. Something has to hold the repeated elements together as a temporal entity. Repetition has to posit a relation to be coherent. In ancient arithmetic, the smallest number is two, because one is their combination.[10] For something to exist, it happens twice. An event is never singular.[11]

Imagine a simple case — if not the simplest. Two beats. This is repetition at its most rudimentary, rhythm at its bare minimum, entrainment in its shortest form. Just a couple of beats, in any tempo, at any perceivable frequency. They may have different properties, but the beats are similar enough to be understood as a repetition comprising two events in time. This simple case is all that is required for coherence.

This is the second premise. Repetition at its minimal limit is already coherent. It posits a relation. Nothing else is needed for music to hold itself together as a discrete duration.

3. Differential Equations What is this relation? This simple case could be described as:

$$A + A$$

or

$$A = A$$

We may assume that the addition (+) or the equivalence (=) results in a coherent connection, but what actually connects two elements together is neither the first "A" nor the second "A," but the unspecified "+" and "=" between them. The binary system involves an "invisible" or "tacit" third element — a relation — that simultaneously joins and separates the two events, like a clip or a fold. This edge or join can be pictured as a line between the repeated units:

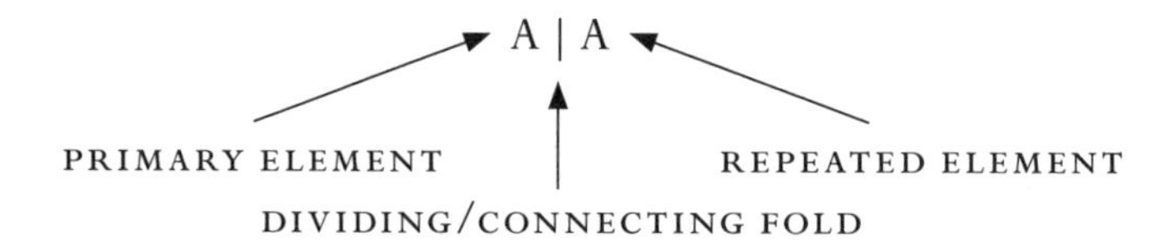

This articulation could be a gap or a noise, but whatever it is, it is fundamentally different from the elements on either side of its position. It does not repeat, but is delineated by the "edges" of the repeated elements. It is both punctuation and transition, division and connection, memory and potential. It is an inherent property of both sides (A and A), accentuated by their moment of contact and yet distinctly different from either side as a moment of reciprocation. This articulation defines repetition. As the turning point, the fold makes repetition possible and determines what follows it, placing the repeated element in relation to the first.

The third premise is that difference is fundamental to repetition. Repetition is not merely about identity or equivalence; it is, already in its simplest form, about difference and disruption. Musical repetition is difference-in-relation. Or to put the matter the other way around, music is nonidentical repetition. Difference is, therefore, inherent in unity.

4. Loopy and Snoopy If difference is inherent in unity, then music is never a single, uniform event. Its unity is always already relational. "One" in itself is never coherent. Even in an extreme case where one event is perceived as a coherent duration (an isolated note in Webern's Symphony op. 21, for example), what we hear as music is not a single event. It is the negative image of the simplest case — "A | A" — where the singular event functions as the point of difference, separating the silences that lie on either side of its position. The silence "A" is repeated, and the note "|" places the point of difference as an aesthetic revelation in a Zen-like state. There are two implications.

First, the inversion places the fold prior to the original "A." This indicates that there is always something before the "beginning" ("A"), making any sense of "origin" random, if not asymptotic. The point of difference "|" is itself already a repeat of a difference that murmurs *before* the "beginning" as a potential to repeat.

(|) A | A

Like a conductor's minimal gesture before the orchestra sounds, time is never purely present at the start, but is already divided and related through difference from the past. So although "A" might be termed an "original" or "primary" unit that is subsequently repeated, it never coincides exactly with itself to constitute a zero origin. There is always an anacrusis...and an anacrusis to the anacrusis, pointing backward to some unrecoverable ground. The same principle applies to the end: it does not end — a pulse is still projected virtually after the final cadence of a piece. The end is not followed by nothing; it keeps going. Music is a knotting together of endless strands of time. It is a piece of infinity.

Second, the inversion indicates that the turning point in the loop can itself be illuminated and expanded. Technically, the point is an impossibility in the same way that a one-dimensional point is a necessary, but impossible aspect of reality. But in the case of an inversion, the point can be pointed out. If the point of difference can be illuminated, then musical repetition can be perceived as a kind of "negative field," similar to the optical illusion in Figure 2.1 where the eye may oscillate between seeing a vase or two faces. The tripartite configuration of any repeated unit has the potential to invert.

Figure 2.1. Vase/face oscillation.

The liminal case in music is John Cage's 4′33″, in which the articulative gap is projected as an environment where random noises from the audience are placed as points of difference that are infinitely variable under chance conditions. Audience noise articulates the "silence" where the edges of the 4′33″ duration are marked by the performance gesture of the pianist (or in the case of the orchestral version, the conductor). It operates in a negative field.

To return to the minimal case of the two beats, the same reversal could occur: the point of difference could "materialize" and create a negative field. Imagine two beats separated by a silence; the articulative gap could be of vastly different durations. It could be considerably longer than the repeated units and could even be "filled in" with other sounds, pushing the singular event in the middle to the forefront of consciousness.

A | A would then be pictured as A ■ A

Repetition suddenly has a middle section. In formal terms, repetition is an ABA structure. And this B section can be anything from a momentary blip to a substantial duration of new material, but in all cases, its uniqueness must be theorized as an integral and essential process that makes repetition coherent.

If the B section is perceived in a negative field, then this unique event could be repeated, turning the ABA configuration into a BAB pattern. But the property of the B section is not the same as that of the A sections. Unlike the A sections, the B section need not repeat: it can be different every time, recurring as a C or D or E section — as in Cage's 4′33″. It paradoxically repeats difference. So although it could (and often does) return as the same articulation, its significance lies in its potential to be utterly different in each manifestation, redefining the relation between patterns of repetition and the expectations of recurrence. Since the point of difference can be infinitely different or similar, it structures anticipation — that is, the play of the similar and dissimilar — and so drives the process of continuation as an open future. It is exploratory, because it is always potentially contingent and undetermined. Repetition, in its simplest formulation, is both loopy and snoopy.

The fourth premise is that the point of difference articulates expansions, disruptions, inversions, and negations within the process of repetition, creating a counterpoint of patterns that are simultaneously continuous and contingent. Repetition is not a teleological, but an emergent process.

5. Free Form In this two-beat scenario, it is not only the "fold" in the doubling process that articulates difference. The repeated beat is itself transformed as the result of "the middle B section." The fold is the immanence of the new.[12] The point of difference makes a difference to the second A so that the same is never exactly the same. The fifth premise is that the repeated unit (the second A) can never be a clone of the primary unit (the first A), but is always already a variant. The two A sections should be redescribed as:

$$A_1 \mid A_2 \text{ (where } A_2 \text{ is a variant of } A_1\text{)}$$

This is not because of their position in time, but because the second A is always a variation of the first A. It is a response. This is a critical difference. The nonidentical element within the repeated unit creates a *relation* that alters the second unit as something dependent on the first, so that it cannot simply be detached as another primary unit. There is an element of entropy at play here: once difference disturbs the process, the second beat becomes more chaotic, more complex, more unstable. Hence, in musical terms, this beat is called the "upbeat" or "offbeat" or "backbeat." Something has switched "direction." What was "on" is now "off." What was "down" is "up." What was in front turns back. The two A sections could therefore be redescribed as:

$$A \mid \forall$$

This reversal registers three properties inherent in repetition.

1. It is a binary system—a blinking on and off.
2. It is a rotational system—a turning back and front.

3. It is a symmetrical system — a mirroring left and right or up and down.

In effect, they are the same system, where the play of difference can be described as the simplest computation or the simplest unit or the simplest reflection. These reciprocal pairings — up and down, on and off, front and back, in and out, or left and right — describe a binary, symmetrical motion that is a rotation or, in linear terms, a wave. Thus, a single rotation, as in Figure 2.2, is never a single component: it is irreducibly binary and is always divided as a symmetrical relation. What goes upbeat must come downbeat. It takes two to be one.

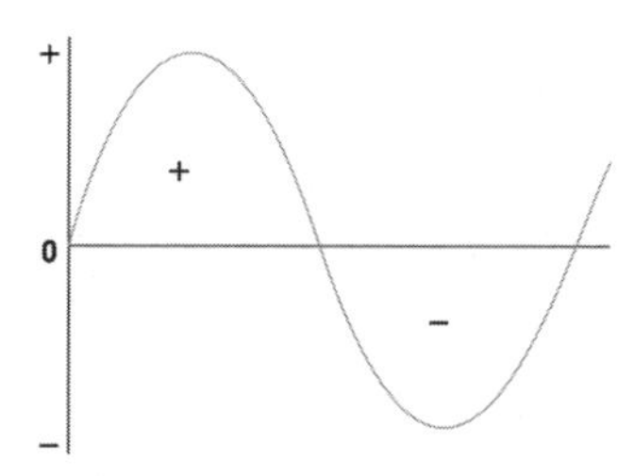

Figure 2.2. A discretized wave producing a binary (+/—) rotational and symmetrical pattern.

This perception is to be expected from bipeds with binaural hearing and a circulatory system that beats to the rhythm of a double pump. Humans move to this rhythm. We rotate. We loop. We shuffle back and forth in physiological cycles. But if everything repeats, then this wobble is not merely human, but is equally enmeshed in the fabric of the universe. The wave is a universal law that wiggles across time and wobbles in things.

So far in the description of the wave, the point of difference has not been figured in, although perhaps it has been tacitly assumed. This point is at the center of a rotation. It is the still point of the turn. Or if we imagine repetition as a wave weaving up and down, what holds the fabric together is the point that runs "perpendicular" to the wave, dividing every point of symmetry with a two-dimensional axis. The point of the difference, in Figure 2.3, is the weft for the warp of the wave; its up-and-down motion is an over-and-under entanglement in a two-dimensional plane. Repetition therefore involves a loom, although, unlike a standard loom, the weft need not be regularly spaced or be of the same material with each iteration. The point of difference is therefore a directional difference, an angle, a moment of tension and intersection that articulates space-time as a piece of fabric. Music is fundamentally interstitial.

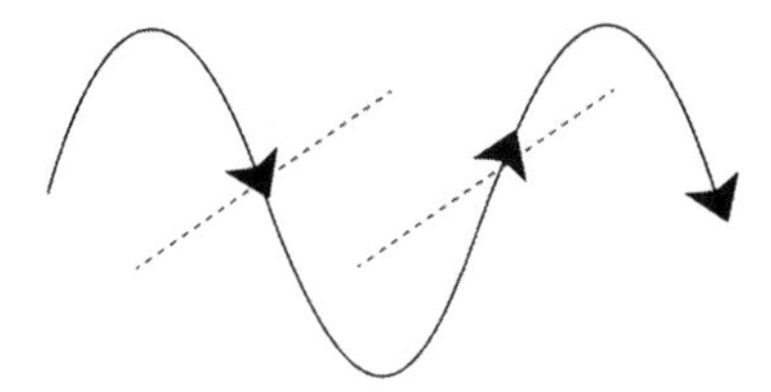

Figure 2.3. A wave "weaving under and over" the point of difference.

One more dimension needs to be factored in: time. All this weaving and turning moves in time. In Figure 2.4, repetition can be described as: 2*D*+*t* (where *t* is time). Time, as represented in this basic diagram, should not be mistaken as a line occupying a preestablished space, but is the very fabric that is being made by the weaving.

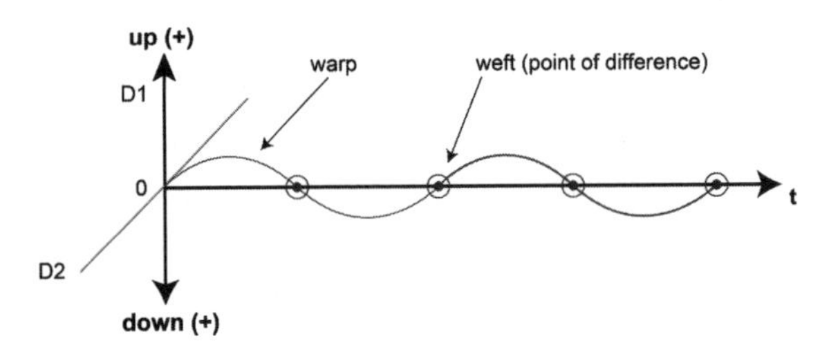

Figure 2.4. A weave in time—2*D*+ *t*.

Since, in accordance with the second law of thermodynamics, time is an arrow that moves in one direction under the condition of increasing entropy, the symmetry can never be an exact mirror. The repeated element (Ɐ) will be different—more complex, more unstable, more divergent. However, unlike the point of difference, it can never be utterly different, since the symmetry is a condition of repetition. Rather, time gives the repeated element a freedom *in relation* to the origin. Repetition is always already a variation, a wave that returns to itself anew to move forward as if to riff off the past to explore the future. Creativity (whether voluntary or unintentional) is therefore a relationship constantly released by repetition as it snakes across time. Dependency does not annul liberty; it generates it in a continually recursive process where the output feeds back as input. Music learns by rote.

Music's generative cycles have two functions. First, they counteract the diminishing returns of entropy; nonidentical repetition, with its permutations and combinations, acts as a counterpoint to the increasing chaos of time. It keeps music going. Second, this relation enables music to anticipate its own completion; the dialectic between creativity and chaos is a negotiation toward a limit within the process of variation itself where the two forces find their moment of annulment: the process is both unpredictable and undetermined, but the continuity of music will always imply its end. Only the identical repeats to infinity; it is both ad infinitum and ad nauseam.[13] In contrast, the nonidentical, for all its multiplicity, is a finitude that creates the possibility for repetition to be *a piece* of time. So although music is a knotting together of *infinite* strands of time, the knot itself is a particular entity—a time capsule. The continuation of the repeated element is therefore not determined as an endless process of replication, but is open to both contingency and closure. The potential to continue the same differently is an irreducible component of repetition—indeed, of coherence itself and therefore the possibility of enclosure.

In summary, music's *2D+t* motion of twisting and turning ensures that repetition is a coherent weaving of time. As weaving, music makes its own surface and its own holes: this porous fabric has no inside or outside, but holds itself in tension as a piece of time.

Repetition, as a process of variation, contains a generative freedom that animates the fabric. Music's flatness ($2D$) in time ($+t$) is sufficient and does not need to be subsumed by a higher order of unity or a deeper principle of form for a coherent play of difference-in-relation. It is not architectural (constructed with building blocks from a plan) but "archi-textural" (woven from lines as an open-work mesh).[14] The coherence of music is therefore an emergent property of these basic principles of repetition, generating both continuity and closure.

6. Digging the Dirt Given this unity, music need not concern itself with coherence. It is fundamental to its simplest process. Repetition is basic and so robust a system that it can withstand continual interference without time falling apart. This premise reorients the direction of music theory, which has invested too much in its search for unity and equalizing difference. To focus on coherence is like listening to the click track instead of the song, or wearing a loom instead of the fabric, or tapping your feet instead of playing the drum. Entrainment, after all, is a given—it is involuntary. What really "counts" is not what can be removed or consumed in the act of making (a click track, a loom, a bit of foot tapping), but the complex, nonlinear "otherness" that is made as a result of what is given, with all its patterns of interference. Music, as a generative repetitive process, focuses on disruption and disturbances—and the dirtier the better.

In accordance with the time-bandwidth theorem, a pure frequency—a perfect sine wave—being edgeless and without boundaries, will always be badly defined in time. A "dirty" frequency, on the other hand, is well defined; it makes time articulate. Dissonance, noise, accents, intrusions, and punctures—such dirty details are the tell-tale signs of time and the very stuff of music. Every additional parameter—pitch, attack, articulation, volume, timbre, text (to name but a few)—superimposes another layer of "dirt" onto the basic process of repetition, generating larger and more convoluted periodic sequences. Together, they corrupt the repetitive cycles until they wobble out of kilter, creating complex textures of loops and folds that weave the fabric of time into discrete, jagged pieces. Their composite rhythms scrunch and stretch and rip and curl the edges of time. Their entanglement generates a surface where ordered differences can plot patterns across the texture—time lines that run "diagonally" or curve against the weave, or pixels that puncture asymmetrical melodies into the fabric.

Dirty rhythms grow by accretion and entanglement, with layers that interpenetrate each other forming infinitely complex cycles to articulate time. They operate like worlds within worlds, each with their own coherence, yet all inextricably bound by a rhythmic viscosity of interconnected permutations and combinations. Paradoxically, the more incalculable the music, the more it delineates a piece of time. And the more it defines time, the more coherent the process. In fact, dirty rhythms are extremely sticky; they do not just cohere within themselves, but adhere to things. They are as gloopy as they are loopy, making time stick to the soles of your feet in a dance or cling within your psyche like a wriggling earworm. The more defined the rhythm, the greater its agglutination. Dirt truly sticks.

The sixth premise claims that repetition needs to be dirty and edgy to define itself in space-time. From the tiniest oscillations to vast rotational forms, music is a pulsating fabric of overlapping tracks, a bottom-up, additive (and addictive) activity that constantly reshapes time with its different riffs and loops.

7. Order of Magnitude Everything in music is repetition. Pitch, rhythm, form—they all cycle along a continuous frequency spectrum. Since music needs to be "dirty," however, it would be a mistake to imagine a smooth sliding scale where everything relates edgelessly across the surface. The sliding scale does not slide. If it did, the "dirt" would not stick to music. Rather, the scale has discrete edges that enable our species to "muck around" with time. Humans process music differently, determined roughly by different orders of magnitude along the frequency spectrum:

- pitch in kilohertz (kHz)
- rhythm in hertz (Hz)
- structure in millihertz (mHz)

This is purely human and peculiarly relative. An elephant, to take a nonhuman, earthbound, floppy-eared example, can hear sounds as low as 15 Hz as some kind of continuous sensation, whereas human ears can experience such a frequency only as rhythm (or pulsation). Similarly, it is conceivable that some extraterrestrial life-form with, say, three ears the size of a Brussels sprout may be able to process fifteen-minute cycles as a rhythmic groove, but humans need to structure this rotation as form in order to hear it repeat.[15] In this order of magnitude, the larger the rotation, the dirtier the frequency needs to be in order for it to be perceived as coherent in time.

These discrete bandwidths that define pitch, rhythm, and form create distinctly different types of repetition. A drone on a sitar (pitch), a funky groove (rhythm), or a baroque ritornello (form),[16] for example, all operate according to the same premises of repetition outlined in this blueprint, but they cannot be reduced to the same process of cognition. Each level has its own speed and temporal jurisdiction and mode of hearing. Their interaction along the sliding scale is a difference-in-relation, enabling music to connect across bandwidths in multilayered cycles that are highly differentiated. But because these discrete bandwidths operate along a single frequency spectrum and work recursively according to the same principles of repetition, there is no power hierarchy. A flat surface provides no privilege. Form, rhythm, and pitch are all independently coherent and freely relate with each other as generative and interactive processes.

At this point, a distinction needs to be made between "rhythm" measured at a particular frequency (Hz) and "rhythm" as a general theory of repetition. The former is found in a specific segment of the frequency spectrum for human ears and interacts equally with pitch and form. It is not fundamental to IMTE. The latter embraces the entire frequency spectrum and is not species specific; it applies to anything that repeats, which is basically everything. This rhythm is a fundamental component of IMTE.

Summary

This blueprint for a theory of repetition is based on number of premises that will need to be tested and developed further:

- Repetition makes "a piece" of time coherent.
- However complex the repetitive process, the simplest case is sufficient for coherence: A + A or A = A.
- Yoking the repeated elements together is an "invisible" fold or clip that simultaneously acts as a divider and join: A|A. This point of difference adds a third element to the twofold pattern, bringing equivalence into relation. Difference-in-relation is fundamental to coherence.
- The point of difference acts as a pattern changer that can be expanded and repeated with the potential to be uniquely different each time. The third element introduces an openness to contingency both within repetition and with its continuation.
- Hard-wired in the repetitive process is a symmetrical motion that connects the repeated element as a variation of the primary element: A|∀. This functions as a binary system in a two-dimensional space under the entropic motion of time. The freedom to continue the same differently is an irreducible component of repetition.
- Music is a process generated by frequency interference. The "dirtier" the interaction between frequencies, the more defined time becomes. Coherence is *simply* given by repetition, unleashing the creative process as an interplay of complex disruptions.
- Interference also functions along the frequency spectrum, dividing a continuous scale into distinct domains of pitch, rhythm, and form. These discrete segments are separated by an order of magnitude (kHz, Hz, mHz), each requiring a different mode of cognition. Pitch, rhythm, and form are therefore distinct and

independent, yet related as processes of repetition along a flat, continuous space, interacting across bandwidths in multilayered cycles of interference.

- *Conclusion* The three components that make repetition possible (A | ∀) forge a relational unity founded on a generative difference. Repetition as the irreducible basis of music gives rise to a potentially endless play of rotational interaction. Since unity is presumed as a basic, nonreducible process, this theory of repetition is paradoxically one of difference and abundance. Everything keeps repeating because nothing repeats exactly.

It might be tempting, at this early stage, to turn this putative theory of repetition into a new kind of formalism and recreate the very thing we critiqued. After all, music is intrinsically repetitive, and it is entirely feasible to tie music up into a tightly bound knot of internal relations that can bounce around in a small corner of academia. In fact, significant insights can be gained by playing such games, since nothing is excluded in IMTE (as stated in Rule 2), but what is intrinsic to music also curves inside out along a Möbius strip: *theoria* turns into *media*. Along this loop, music is as boundless as it is bounded, and the purpose of IMTE is to embed music everywhere without unraveling its identity (Rule 3). It is as formal as it is cultural. If everything repeats, then music will have a spin on everything. By keeping the theory simple (Rule 1), IMTE is designed to attract an eclectic and chaotic subset of everything so that any discipline can participate in its theory. Our theory of everything is modular, because it can be gathered together only from different points of view.[17]

0101. WEAVING

Rhythm is the irreducible basis of music. Although Pythagoras was right about the universe as a kind of musical vibration, he was wrong to attribute music's fundamental structure to the harmonic series of a single string. Unfortunately, music theory has ended up disregarding what was right and has developed what was wrong. The rule of harmony has ring-fenced much of music theory. So it is at this point that IMTE departs from Pythagoras, because tones, let alone the harmonic series, are a later addition to music's primal oscillations. Music is fundamentally rhythm. And a theory of repetition is fundamentally rhythmic and should be separated from the ideological consequences of Pythagorean thought.

A harmonic cosmos is immovable and total. Although its fundamental vibration wields absolute power, its iron grip belies a fundamental insecurity, for its mathematical proportions are so perfect that they harden into an inflexible universe where the tiniest hairline fracture could cause the entire edifice to collapse. There is simply no give. The music of the spheres emanating from a hard, crystalline surface is prone to crack. In many ways, the history of Western music theory is an elaborate attempt to paper over the crack to stall the cosmic disintegration. A rhythmic cosmos, on the other hand, is flat and pliable, because its patterns do not form stratified structures, but interpenetrate each other in their difference to weave a translucent, relational, unfolding surface. Instead of a totalizing structure, it begins with tiny dots—blinks, beats, gaps, binary patterns. From these dots, lines come into being and bend into waves as they weave in and out of each other. A dot should not be conceived as a bit of a line, but is its generative basis, as Leibniz might explain, resulting in a crisscrossing of fibers spinning from each point to create a mesh of frequencies.[18] There is no inside or outside; there is no hierarchy; there is no conquering or dividing of a preexistent territory; there is no possession. The surface of time and space simply emerges as a weaving of knots and holes.

Hence in the theory of repetition, music is best described as a "contrapuntal" fabric. Its iterative linearity allows for a transmission and transformation of information. Its repetitive rhythms weave time in a flexible and infinitely extendable two-dimensional space. As a weaving of time, music is not held together by some Pythagorean totality; it neither emanates from a fundamental base/bass nor descends from some transcendental unity. Rather, music is a network or meshwork that spools and sprawls without a center or a fixed boundary to control its movement. There is no single string, but, instead, there are many strings woven in relation. This weave is manufactured, rather than metaphysical, as algorithmic as it is rhythmic and as mechanical as it is malleable. It is capacious, thriving on alien disruption to tighten its coherence. It is generous. It gives. It flexes. It is virtually indestructible. As a fabric, music can crumple, twist, stretch, or fold to conjure up the semblance of a three-dimensional space; it can even assume a "solid" shape, as in basketry, but this illusion should not be mistaken as an underlying structure that contains or supports its motion. It has no depth.

Neither is repetition a reductive form of knowledge, bent on distilling an essence. Instead, repetition is a process of addition that proliferates and generates; every addition, however small, changes the feel and fold of the fabric. This process is about dimensions, rather than centers, distribution, rather than hierarchies, textures, rather than structure, patterns, rather than form, edges, rather than endings, knots, rather than blocks, rhythm, rather than tone.

If music theory were to have an origin in ancient Greece, it would not be Pythagoras, but Penelope who would inspire our theory. In Homer's *Odyssey*, Penelope weaves and unweaves a funeral shroud for her father-in-law to keep her suitors from claiming her during the long years while her husband, Odysseus, is away on his epic adventures. Her weaving is the holding pattern of her moral universe, a *fabric*-ation for shaping, delaying, and anticipating time through a rhythm that waits for the ultimate cycle to return. She is, in a sense, weaving time as her moral fiber, enchanting the world through a technology of binary computation to manufacture a piece that can be done or undone at any time.

A Penelopean music theory, as opposed to a Pythagorean one, would be a purposeful, yet endlessly open process of repetition in time. Her weaving, in its doing and undoing, is open to the future; indeed, it is faithful to the future in keeping it open against the oppression of the present. Its ways of enchantment, as in traditional cut along the dotted line and roll page to make your own three-dimensional Grecian urn of Penelope at her loom

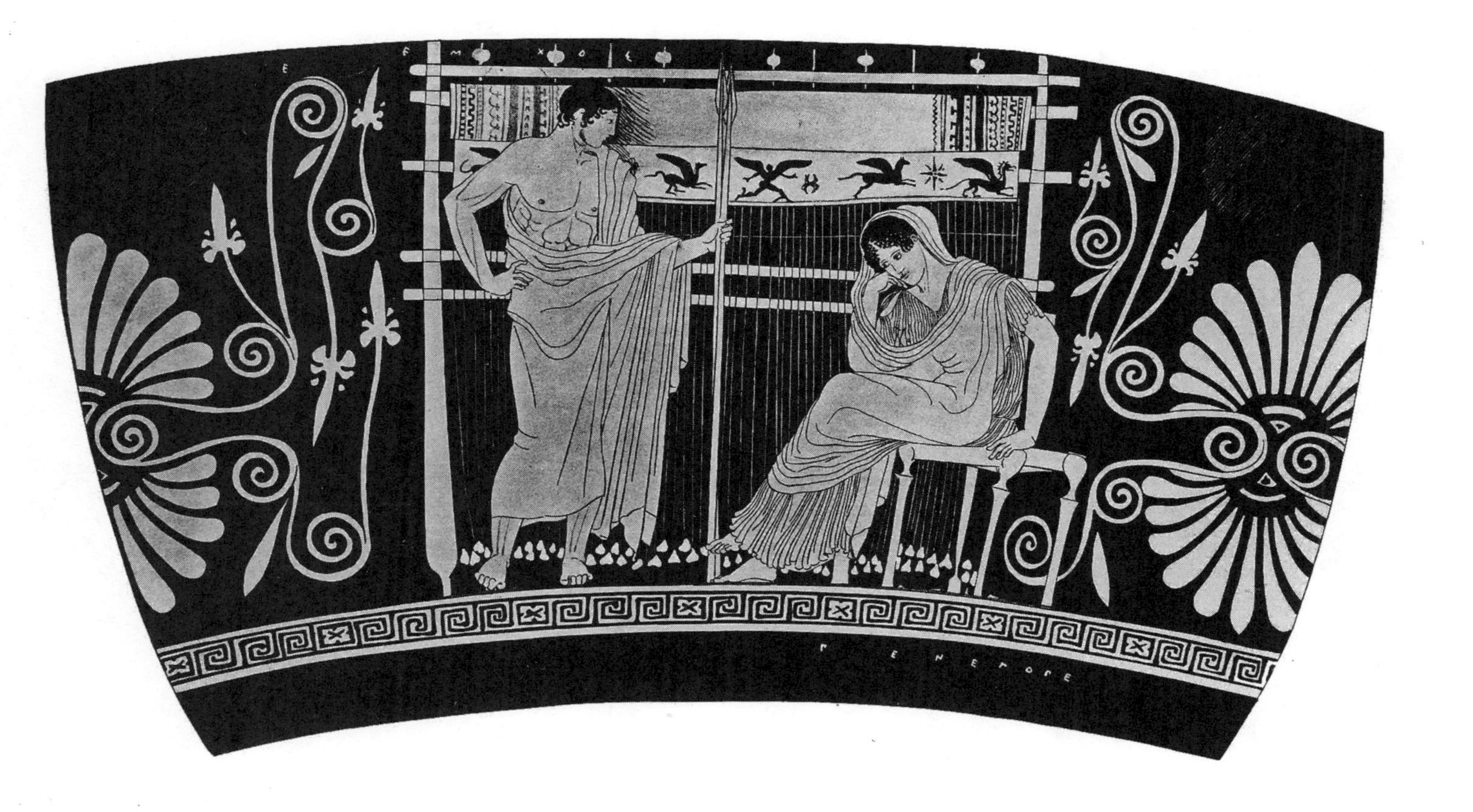

Penelope taking a rest from her work at the loom, from an Attic drinking cup, ca. 440 BCE. Museo Nazionale Chiusi (image: Alamy).

PENELOPE AS MUSIC THEORIST

The tale of Penelope's shroud is told no fewer than three times in the *Odyssey*, in books 2, 19, and 24. It is aptly a weave in the text, a rhythmic fold of beginning-middle-and-end. In both Homer's *Iliad* and *Odyssey*, weaving does not only produce textile; it produces text. Homer's own handiwork is the weave of poetry. As Kathryn Sullivan Kruger suggests, in the *Iliad*, Homer identifies with Helen of Troy, who weaves war heroes into her tapestry to bestow immortality on the warriors of the Trojan War in the same way as Homer immortalizes them in his poetry: "Helen's shuttle is analogous to Homer's voice."[1] Penelope's weaving is different. In the *Odyssey*, she weaves a web of intrigue rather than one of poetry, since her act of unweaving ultimately produces a blank sign. Her "text" is perpetually undone. Penelope does not spin a narrative: she is weaving time, producing a rhythm on her loom that empties the narrative content of time to stall her own story from becoming a tragedy. She weaves music.

Penelope points to a tradition in which weaving and music were intertwined as a computational technology in Indo-European cultures long before Pythagoras crystalized the universe into musical spheres.[2] Women sang as they wove, and the loom functioned as a kind of lyre for their song. But the purpose of these songs was not to weave a heroic tale; it was less a text than a texture of counting, for the rhythmic mechanism of the loom was controlled by a form of sonic software for pattern making. These songs were repetitive numerical codes used as mnemonic devices to program weavers to count complex combinatorial sequences and transmit designs that snaked like melodies across the woven fabric.

Since Penelope's fabric was intended as a funeral shroud for Laertes, it is likely that her weave would be highly elaborate, requiring a song to program the intricate patterns. So, one could imagine her singing as she wove (or at least whispering a song under her breath, given her clandestine weaving).[3] If so, then Penelope's shuttle is analogous to her song. Her woven music was simultaneously temporal and spatial, performed in time and recorded in space. This was a music that blinked in binary patterns. In this sense, textiles were the first format for storing digital music data.

1. See Kathryn Sullivan Kruger, *Weaving the Word: The Metaphorics of Weaving and Female Textual Production* (Selinsgrove, PA: Susquehanna University Press, 2001), p. 78.

2. This musical warp and weft is part of a broader musical cosmology. As the Sanskrit collection *Rig Veda* 10:130.2 explains: "The Man stretches the warp and draws the weft; the Man has spread it out upon this dome of the sky. These are the pegs, that are fastened in place; they made the melodies into the shuttles for weaving." *Rig Veda* 10:130.2; Wendy Doniger O'Flaherty, trans., *The Rig Veda* (London: Penguin Classics, 1981), p. 33.

3. See Anthony Tuck, "Singing the Rug: Patterned Textiles and the Origins of Indo-European Metrical Poetry," *American Journal of Archaeology*, 110.4 (2006), pp. 539–50.

cultures and mythologies of weaving, are the preserve of women and goddesses, offering a computational, mechanical, manufacturing process that is feminine, rather than masculine. There are no male heroics here. Indeed, it outwits their grip. Instead of exercising conceptual control over nature, a Penelopean music theory is a practice that waits in accordance with the rhythm of time, attentive to the cycles of the universe, hoping beyond hope for an encounter with a stranger with whom she is somehow familiar.[19] Penelope models the hopelessly optimistic vision of Voyager with a music that faithfully waits, perhaps forever, to be heard again.

Odysseus returns home
Big surprise! I got you the new
Loomatic Web-Plus from duty free.

In this sense, music is fundamentally a lavish waste of time and resources, because it is ultimately a symbol of high fidelity, representing the faithfulness of our species to relate to another, to be remembered by another, and to outlast our ephemeral flicker in another. NASA's mission is as much a memorial (a funereal shroud) as it is a forestalling of the present in the hope of future recognition.

But perhaps the most compelling parallel is the fragility of the whole enterprise. A Penelopean theory of music begins with a loom—an ancient computer, with a binary grid that switches on and off, shuttles back and forth, and rattles up and down. It is more a thing than a string, too rigid to model the flux of time, but what begins on a machine ends up as a fabric so ephemeral that it is unlikely to outlast the machine on which it was made: music is a vulnerable intelligence that speaks of posthuman fragility. What is discrete in time vanishes in time.

The history of notation and the technology of sound reproduction describe our attempt to capture and sustain music's ephemeral nature, to reattach the fabric to a machine so that its repetitive patterns may continue to recycle themselves forever. Like Penelope's shroud, music's relationship to its medium and machinery is a recursive tic that binds them like a mise-en-abyme. It is as if this relationship functions as an artificial life support that keeps repetition breathing; music lives as long as its data are repeatable. It switches on and off as event and data storage. A Penelopean theory of music, unlike Pythagoras's immovable ratios, recognizes music as a prosthesis, dependent on a technology of manufacturing and reproduction, not because music is eternal, but because music is temporal and material and is always already being undone the moment it arrives and must find ways of repeating itself.

Summary IMTE, despite its indebtedness to a Pythagorean vision of music theory, operates in opposition to many of its ideological premises. To reimagine music as a weave or fabric, analogous to Penelope's shroud, is to invite us to rethink theory on an ontological level, enmeshed in an alternative texture and attached to a machinery of repetition. "Perhaps, the highest object of Art is to bring into play simultaneously all these repetitions, with their differences in kind and rhythm, their respective displacements and disguises, their divergence and decenterings. . . . Art does not imitate, above all because it repeats; it repeats all the repetitions, by virtue of an internal power."[20]

0110. INSTRUMENTAL MUSIC

Year: 2020 — as we write, Voyager has been barreling through space for well over forty years. Yet on its way to the future,

Voyager's golden LP is already a relic of the past. Shipped off to space before music's digital revolution, the disc is now hopelessly out of date. It seems that forty years, let alone forty thousand, is enough to render its technology ancient. But strangely, this quaint analog mechanism communicates in far more concrete terms than any form of digital data, which is too virtual to grasp. If an alien were to intercept the space module, it would not discover music, but "frequency-making" objects—the rudiments of a turntable, a rotating disc etched with ridges and grooves, a stylus designed to vibrate, and a set of instructions with "hieroglyphics" (in Figure 1.3) for setting the right frequency for playback. These are artifacts intended to invite the alien "other" to resonate with us, like a golden handshake undulating across the stars. Human music is being delivered to the universe as a "thing." NASA's disc is as much a matter of "thing theory" (a redundant object reused by aliens) as it is a string theory (information as vibrating cycles)—a kind of dis-located frequency machine.[21]

Any extraterrestrial life-form that intercepts Voyager will have to read humanity from this frequency machine, just as any alien who discovers our planet would have to deduce our musical identity from the fossilized remains of technology embedded in the Anthropocene. This is apt, since human "musicking" is defined by technology, as Gary Tomlinson points out. There is a thinginess that externalizes our music *making*. Even presapiens hominids, for Tomlinson, composed a "taskscape" of ambient sounds with the repetitive motion of their flint-knapping activity: whatever these tools were used for, instrumental music was a by-product of these instrumental artifacts: the body-stone interaction of hominids functions like a collective "rehearsal machine," with cognitive feedback loops refining their rhythmic synchrony to compose the incidental music of the Paleolithic.[22]

A million years later, this "rehearsal machine" has evolved into the frequency machine on Voyager; the golden disc with its stylus is an objectification of the cognitive feedback loop in the form of a prosthetic extension and technical detachment of human musicking.

The evolutionary timescale from the earliest "rock music" of the Paleolithic era to the rock 'n' roll of the Chuck Berryian era has been compressed into its rotating grooves as a testimony to the technological materiality that defines human music.

Hey, Gary, I think we've just invented rock music.

At first, it may appear that such instrumental objectification would estrange music from nature. Instrumental reason has often been accused of being the Midas touch of humanity, alienating everything it grasps in its quest for knowledge.[23] After all, what else can this frequency machine be but an object turned to gold by the touch of NASA's rationalizing finger, striking music dumb and cold in the process? Floating in interstellar space, music has been literally instrumentalized by human intelligence, seemingly disembedded from the environment and stripped bare of cultural meaning.

But the opposite is also true: far from estranging us from nature, musical instruments disclose our embeddedness in it. Music technology is not simply a prosthetic extension that alienates us as tools, but an externalization or amplification of self-knowledge that reflects back a nature so integral to our being that it is often hidden. When Herder imagines the clavichord as a mechanism "within us," he in effect turns our auditory system inside out, as if the wooden instrument were a physiological revelation of an inward audition: the clavichord demonstrates how our inner frequencies tingle and react to the vibrations around us. Similarly, when Diderot hears the nerve fibers of the body jangling like the strings of a harpsichord or when Helmholtz models the cochlea as a piano, as in Figure 2.5, they demonstrate how science has always made man in the image of sound technology.[24] Musical instruments are a kind of exoskeleton of an inner anatomy, a prosthetic reimagination of the living frequencies that circulate invisibly within our bodies and animate our minds and how they vibrate with the environment. These instruments are prosthetic in that they are bionic receptors and receivers that enable us to amplify and concentrate the rhythm of the universe for our ears to understand. After all, what else is Pythagoras's "string theory" than a giant monochord unveiling the secret frequencies of the universe both inside us and beyond us? Music technology is the second nature of our ec(h)osystem.

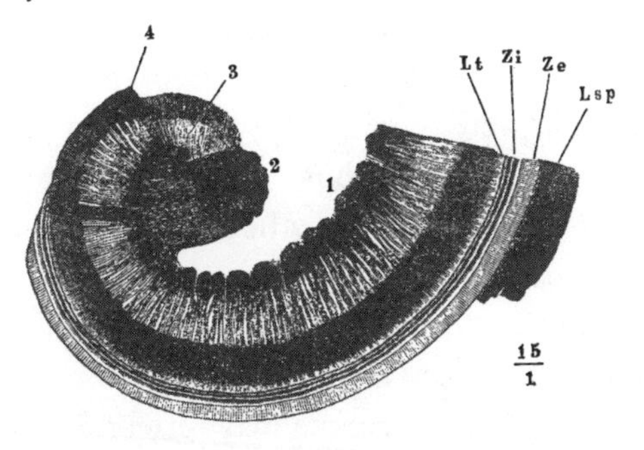

Figure 2.5. For Hermann von Helmholtz, the cochlea operates as if it were a spiral piano keyboard. From *Die Lehre von den Tonvorstellungen* (1863).

NASA's frequency machine, then, is a specimen of human audition that discloses to the alien our obsession with repetition as a mode of perception. We are loopy beings in a loopy world, a species addicted to good vibrations. If an extraterrestrial managed to reinvent the gramophone from NASA's instructions, the music ze "hears" (the gender-neutral pronoun again seems appropriate here) would open a window into our physiological and psychological particularity encoded in cycles. It would be a tactile, vibratory interaction with a makeshift turntable that literally models music as rotation. As our music begins to resonate, the alien would sense how we perceive, manipulate, and modulate a narrow spectrum of frequencies; ze would encounter our planet as if through our ears, in music made for human audition, but now strangely recycled: ze would sense the way we inhabit the fabric of time and the rhythm of life. Ze would receive a limited, but intense resonance of our blue planet, made, mediated, and objectified through technology. Humanity would appear as the ghost in the machine. Given the probability of human extinction within the next million years or so, this mechanical apparition may be all that remains of human music in the future, and yet, if the alien got into the groove, ze would, in a bizarre sense, get into us.

Music theory, then, is not so much about technique as technology. *Techne* — as Heidegger redefined it — is a disclosure of humanity. It is about how we *make* sense. Human music is embodied practice sustained and extended by the machineries of repetition. However, for all its audiophile pretensions, there can be no high fidelity in outer space for NASA's frequency machine. There is only a high fidelity — a faithfulness — in our relation to an other, not a high fidelity to sound reproduction. The music on the Golden Record exists regardless of percipient, but for it to communicate, it can connect only as a differential relation. As we have noted, the music we hear is for our ears only. Voyager's frequency machine may reflect our sonorous self, yet NASA provided no speakers for the alien, inscribing a certain inoperability to its mission, for speakers mirror our ears. By delivering a strangely "voiceless" hi-fi, NASA forces the extraterrestrial to extract the vibrations for zirself. The suspension of sound therefore creates the possibility of meaning, a subtraction that inclines the alien ear to literally *make* sense of sound.

With such *pre*amplification as a condition, there can be no such thing as passive listening in an exoplanetary music theory. In a posthuman pro-alien theory, music is suspended without "speakers" to enable a relation to take place along a scale dependent on differences in physiological, cognitive, and environmental factors that demand an active construction of hearing. Its communication is always particular and very peculiar. Indeed, the alien will not be able to distinguish our music from the other sounds etched on NASA's disc — thunder, surf, wind, birds, whales: it would all be a play of frequencies. Our intentional music would simply be embedded in the accidental sounds of nature that predate us. All the fuss about setting the right frequency for playback inscribed on Voyager's LP will not communicate the music more accurately to the alien: it will just tell the alien how the human ear works. Our music will be bereft of all the meaning that musicology wants to endow on music to separate culture from nature. But in space, it will have no history, no cultural specificity, no tradition, no theory, no context. The alien will hear us in the same way we hear the whale on the Golden Record — as a translation of a nature that we don't understand, but imagine as song.

However, this cetaceous sound is not simply constructed arbitrarily by humans as song; the meaning we make from its strangeness is a property of that sound, a possibility of meaning emanating from within the object that connects one sound world to another. Similarly, estrangement may be the mode of reception in space, yet it will trigger an invitation to imagine us, to understand us, to hear us. NASA's disc, like all obsolete things, is prone to being reused, upcycled, and unraveled. It will carry its own temporality, make its own history, and be its own exception as it moves out of sync with humanity's pendulum. It will not recount our narratives to the alien; it can only count it in frequencies. But this strange thing, this prosthesis of music, for all its redundancy and contingency, will invite a close encounter that may reveal more to the other than what we could ever hear or imagine of ourselves. Like the hammers that caught Pythagoras's ear, the Golden Record may open a new universe for its recipients.

In the meanwhile, between now and the distant possibility of a close encounter, our music is locked in cold storage, waiting for someone to break its silence and release an explosion of frequencies. It is seemingly a game of chance. And yet, precisely because of the improbability of an encounter, it is more a game of patience that questions chance. Like Penelope, this waiting—this high fidelity to a universe predicated on hospitality—is humanity's ultimate act of listening. The universe is not mute; if it were merely a meaningless void, our species would never have responded to its call. Someone is out there. And because the call is anonymous, NASA's strategic mission has drifted out of its normal orbit; Voyager has become a wayfaring vessel in search of the caller, meandering in deep space without a map or a destination. Thus, as a response to an alien other, Voyager is not only "our voice"; it is our prosthetic ear, a *probe* attending to the universe, seeking to interact with the frequency of another. In this long interim, we can but stop speaking in order to take an ascetic journey of silence, because we believe that other places and creatures, however distant and strange, are never mute and must be heard. Our music, curled up in hibernation, is waiting. Silence is golden; music has finally taught humanity to listen.

FOLDING WAVES

Takahiro Kurashima

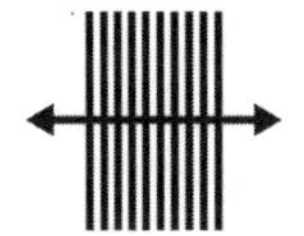

Use the film filter provided with this book (or copy the filter pattern in the Appendix onto transparent film).

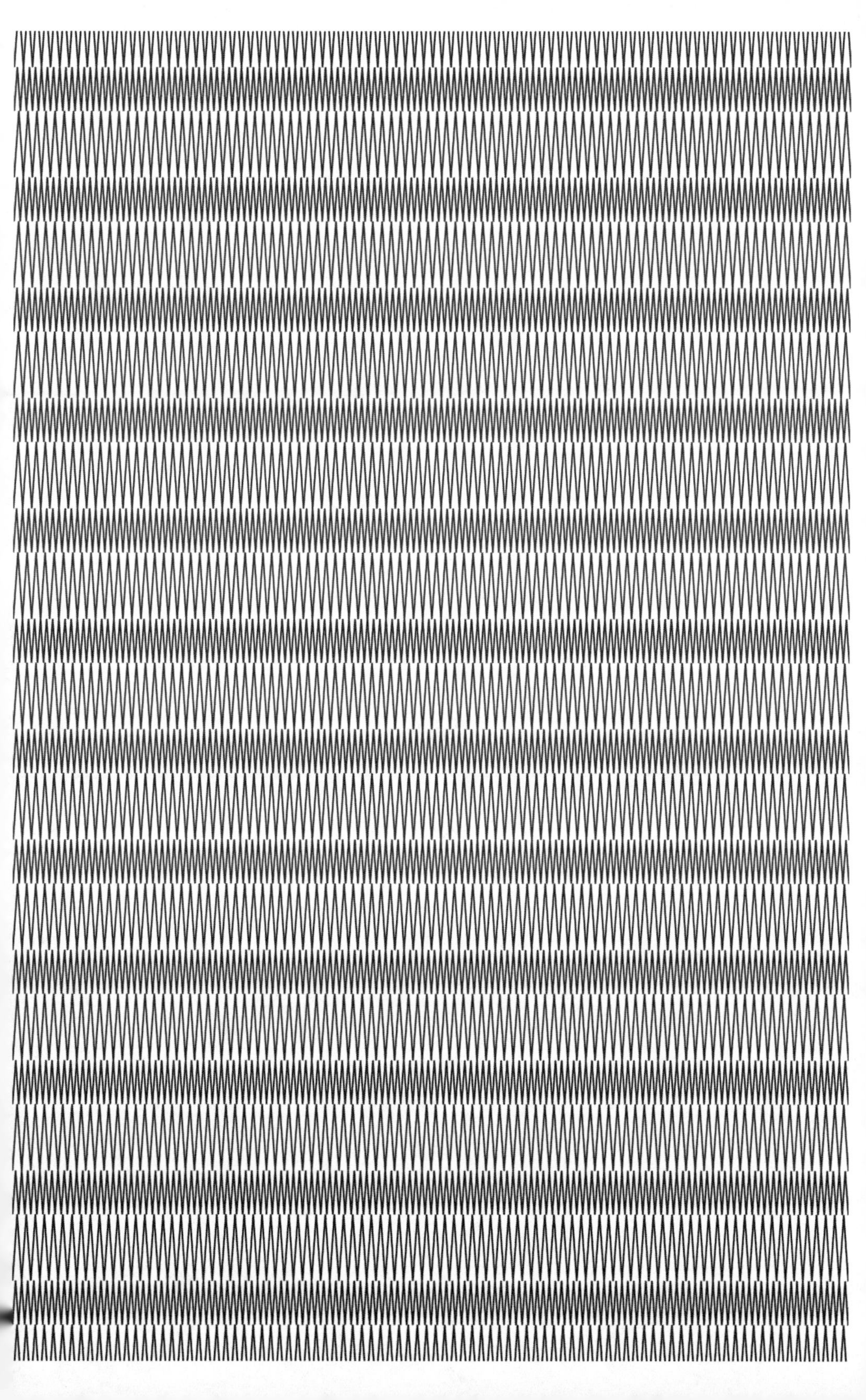

PART TWO

A Media Theory of the Third Kind

CHAPTER THREE

Sender

—John Williams, *Close Encounters of the Third Kind*

It is time to adjust our space telescope. In Part 1, we zoomed out to sketch an intergalactic music theory of everything. The vision was bold and brash, connecting multiple dots across the canopy of space with the broadest of theoretic strokes. Voyager was merely an occasional blink guiding our peripheral vision. In Part 2, we zoom in on a particular object in order to complement a music theory of everything with a media archaeology of just one thing—NASA's Golden Record. The medium is now the focus, and Voyager takes center stage.

The Golden Record projects a sonic vision of Earth. Its glistening surface beckons the alien to listen to our planet and audit a slice of human culture. But what does listening mean in an interstellar context? If music can bring alien species in counterpoint across vast distances, then the question posed by the Voyager mission would push the boundaries of musicology beyond music as we know it; the scale of music would expand exponentially in terms of time and space, and the human monopoly on its significance would fade like a distant planet. An exoplanetary vision demands an exomusicology in which music is reconceived as something prior to *Homo sapiens* and that will likely exist as a posthuman artifact wandering the universe long after our extinction. To travel with Voyager, then, is to enlarge music's definitions radically and to rethink what music is in terms of its theory, its media, its practice, and its sound.

The idea of reimagining music outside our species and beyond our timescale may seem a vastly ambitious proposition with universal pretensions that verge on the abstract, if not the abstruse. But this is not the case, and the reason is simple: NASA is a space agency, and not a department of philosophy. *They make things*—solid, functional objects. Any abstract speculation emanating from the space agency is a by-product of the nitty-gritty materiality of their technological know-how. However ambitious the Voyager proposition may sound, the reality is concrete and relatively primitive. What NASA launched for its intergalactic mission was an analog LP with the rudiments of a turntable, stripped down to its essential stylus and cartridge unit, complete with a diagram for mechanical sound reproduction. Exomusicology begins with pretty basic stuff. NASA's extraordinary mission is ultimately an ordinary task, with mundane things doing what these things normally do. As with all music technology, the main functions of Voyager's bare-bones hi-fi are simply to time travel and space shift. It just does so on a massive scale—it receives, stores, and retrieves data for "all worlds" and "all times," as announced on the record's takeout grooves as seen in Figure 3.1 (if we look very carefully).

If the medium is the message, then what constitutes music on Voyager are precisely the technological artifacts onboard the spacecraft. These objects should not be regarded as extramusical appendages. *Music is always materially bound to its media; it has no communication apart from technology.* As with Pythagoras's hammers in the blacksmith's yard, the medium forges music. Technology, of course, need not be a machine; any sounding body can operate and mediate music—a larynx, an erhu, a gramophone. In the case of NASA's alien recipient, our music is literally a UFO—an unidentified flying object. In our human attempt at exoplanetary communication, it is not so much NASA as this opaque, decontextualized thing that is the real space *agency* here. Humans are ultimately redundant, which means that by the time the Golden Record arrives on another planet and lands in the lap of some extraterrestrial intelligence tens of thousands of years from now, NASA's mechanical music machine would require the skills of an alien well versed in media archaeology.[1]

Figure 3.1. Image of the Golden Record. Tim Ferris, the producer of the Golden Record, etched by hand an unofficial message in the blank space of the takeout groove: "TO THE MAKERS OF MUSIC—ALL WORLDS, ALL TIME" (image: NASA/JPL).

An archaeology is necessary because the Golden Record is locked in its own Anthropocene. Fossilized in a stratum of time, it cannot evolve into higher forms of technology and will someday be forgotten by our species and abandoned without memory as a relic. Cultural meaning will be erased. The endless proliferation of signs and references that light up the networks of human music will fade into the dark, and our very understanding of the sounds that NASA has sent beyond our solar system will be left behind on our planet. The Golden Record is destined for oblivion, losing all connection with its earthly networks. If it is found on another planet, what emerges from the device will not be history, but new knowledge; the Golden Record will have to reinvent itself (or rather re-event itself) for an other. And the alien, in listening to an obsolete past with no historical coordinates, would need to withdraw into the object to refigure the medium. Earth music would be gramophony remastered. What normally effaces itself to transmit music as an acousmatic phenomenon becomes an opaque and obstinate medium through which music emerges less from a flow of data than from its material impedance. It is precisely through an archaeology of this obdurate object meandering in a flat ontological space that a reimagination of music outside our species and beyond our timescale begins. A new network may arise, but not as we know it.

An ancient machine is therefore the focus of our exomusicological exploration to answer the question posed by NASA's musical expedition: Can music be used to communicate across civilizations, across species, and across galaxies? What this answer will sound like, however, is as uncertain as Voyager's mission. In space, music is anything but normal. All we can promise in our close encounter with Voyager is a media mystery tour. We will, of course, contextualize the recording and coordinate the details and narratives that define the Golden Record. After all, our fascination with this object is fueled by the facts, figures, fantasies, texts, ideas, theories, networks, discourses, persons, and things that spin brightly off the disc like a spiral galaxy. But ultimately, to take the Golden Record seriously is to step into the unknown. The stars fade, and we are in the dark. Cultural mediation will cease; social relations will dim. But what we will know for sure is that an imagined encounter with extraterrestrial intelligence will estrange what we think we know and alienate how we think we hear. It will challenge us to cross the final frontier.

0001. AN INTERSTELLAR MIXTAPE

Music has always been a peripatetic companion, navigating vast distances along the song lines of the Australian landscape, booming from car stereos on the streets of Kinshasa, throbbing noiselessly in the earbuds of commuters in Tokyo. Music travels. NASA's conceptual model for portable music was a familiar one for its time: the Golden Record is an interstellar mixtape, compiled from Earth's favorite soundtracks. The space agency did what everyone did in the 1970s to personalize their music, except that the person represented by this mixtape is all of humanity. Flying at a speed of thirty-eight thousand miles per hour, it is on its way to another star system in search of intelligent life in order to share our playlist. We won't know if it will stumble upon a viable exoplanet for tens of thousands of years—plus whatever time it takes to bring this knowledge back to Earth. So don't hold your breath.

The interstellar mixtape left Earth's orbit from the Kennedy Space Center at Cape Canaveral in Florida. Even though we usually speak of the Golden Record in the singular, there are in fact two identical Voyager spacecraft, each with its own Golden Record on board: NASA produced a backup, Voyager 2, in case the first launch failed. To confuse matters further, the second Voyager was sent up first, on August 20, 1977. With the success of this trial run, Voyager 1 joined its twin sixteen days later, on September 5, now with more fanfare and official imprimatur, on their journeys into the unknown.

When the Voyager mission was put together, the Golden Record was almost an afterthought, compiled hastily in six weeks.[2] It was the brainchild of Carl Sagan, the starry-eyed American popularizer of space and science. It was also a child of its time, conceived on a universal high, heady with lingering ideals of the Summer of Love. Among a series of greetings in fifty-five languages, a collection of Earth sounds, and 115 images of our planet, the record contains a compilation of World Music, long before the term became common usage, from Peruvian panpipes to Japanese shakuhachi.[3]

At the time, the idea that there might be intelligent life out there was considered a crackpot theory, the stuff of science fiction pipe dreams. Sagan called it a "semirespectable" idea, which is a polite assessment.[4] In the absence of any material evidence, only a minority of hard-nosed scientists took the idea seriously.[5] Meanwhile, over the last decades, an overwhelming number of exoplanets have been discovered. A few thousand planets have been confirmed by NASA, while the overall estimate currently stands at one hundred billion. The number of exoplanets in the universe is so vast that the tables have turned: to put it with a double negative, it now seems statistically *improbable* that *none* of these planets would harbor life.[6]

This shift in thinking has also changed the mission after the fact. Where the makers of the Golden Record were initially at pains to reassure its critics that the inclusion of the Golden Record should be taken as predominantly symbolic,[7] there is now a nonzero chance (however slim it may be) that the record might be intercepted at the other end and played back sometime in the distant future. As such, the Golden Record doubles as a time capsule, forever preserving a slice of Earthbound musical culture circa 1977 as it zooms away from our planet.

To be sure, the "symbolic" explanation had always been a ruse, loudly proclaimed so as not to endanger the scientific credibility of the mission by giving in to tin-foil hattery. After all, Sagan was a founding member of the Order of the Dolphin (1961), a group of highly decorated scientists who dedicated themselves to the search for extraterrestrial intelligence—it was to become the Search for Extraterrestrial Intelligence Institute, SETI—and he added his name as a coauthor to the English translation of *Intelligent Life in the Universe* (1966).[8] The official scientific purpose of the Voyager mission was to take high-quality photographs of the planets of our solar system for as long as technically possible; its images of Jupiter and the outer planets have revolutionized our understanding of the Jovian planets, and the flybys of Uranus and Neptune, which were not part of the original plan, are now hailed as key scientific contributions. Meanwhile, the makers of the Golden

Record harbored the secret hope that extraterrestrials would listen to the LP attentively—and even apply a little music analysis in the process by comparing works by the same composer on the disc.[9]

The Golden Record has always been a supplement of the Voyager mission—in a strictly Derridean sense. Apparently extrinsic to the mission and its scientific goals, this afterthought emerged as the defining feature that not only completes the mission, but makes it possible. The Voyager mission has captured the attention of the public more than any other space mission with the exception of the first moon landing. By way of comparison, the mementos on board the more recent New Horizons spacecraft, launched by NASA in 2006, contained a postage stamp of Pluto, an American flag, special twenty-five-cent coins bearing symbols of the states of Maryland and Florida, as well as a piece of SpaceShipOne, the first commercial spacecraft to transport people into space. "Which is why," poet Michael Anthony Morena concludes, "compared to Voyager, no one gives a shit about New Horizons."[10]

Part of the Golden Record's broad appeal stems from its ability to bring the abstractions of the universe down to a concrete level, alluding, in a familiar object, to the transcendent idea of music as the universal language. Its message is both everyday and exoplanetary, simultaneously mundane and mind-blowing. But who exactly is this mixtape for? For some other alien life-form? Or is it for us? For the mundane or the exoplanetary? Like all mixtapes, the Golden Record says as much about its maker as it says about its recipients. At once deeply personal and revealing, a mixtape is a record of the living dead: it inscribes who we were at the time we made it and may no longer be. Revisiting old mixtapes is always a nostalgic activity that allows us to reconnect with a past that no longer exists. Every mixtape is a kind of "I will have been," the grammatical tense of the future perfect that inscribes one moment in time into the perpetual present of the medium.[11] Given the extreme time spans involved in the Voyager mission, this is almost certainly true for the Golden Record as a collective mixtape from humanity. The mission, by extending time in all its convoluted retentions and protensions on an unimaginable scale, amplifies the anxieties of our species as a mere blip in the universe. Over the next forty thousand years or more, humankind will be completely different, if indeed it still exists at all. NASA's recording is bound to become a collective relic, because Sagan's team (including chiefly Frank Drake, Ann Druyan, Tim Ferris, Jon Lomberg, and Linda Salzman Sagan) has built into the mission what Evander Price calls a "future monument."[12]

In the distant future, 1977 may prove to be the most important year in human history, if not the only year recorded of our history. As President Carter's typed message on board Voyager intimates, if our struggling humanity proves incapable of solving its problems, the golden disc will be Earth's final testament to the universe. Somehow, the hopes and fears of the Cold War, pressed into a tiny metal disc, speak for the future of all humanity: "We are attempting to survive our time so that we can live into yours."[13] In this sense, NASA's future monument is simultaneously a "late work," a stance poignantly suggested by the curators of the Golden Record by placing Beethoven's Cavatina, from the B♭ Major String Quartet op. 130, as both the final track of music and the final photograph in the collection of images, as if this is the one piece the future *must* understand. This movement, from one of Beethoven's late quartets, carries the weight of lateness that the spacecraft will one day inherit. Its future-perfect statement is not only out of its time, but may soon be too late in time: "What could have been is not to be." And yet in its solitary journey into the night, the Cavatina carries a "little thing of hope," as if it somehow preserved a secret joy, kindling the memories of a forgotten world.[14] The Cavatina is "an appropriate conclusion to the Voyager record" writes Timothy Ferris, the producer of the recording team. "We who are living the drama of human life on Earth do not know what measure of sadness or hope is appropriate to our existence. . . . The dying Beethoven had no answers to these questions. . . . In the Cavatina, he invites us to stare that situation in the face."[15] So for all the claims of scientific progress and the Utopian vision of a "single global civilization,"[16] Voyager's mission takes the form of projected nostalgia: a future monument to teach us to love what we've lost.[17]

In one of his aphorisms, Friedrich Nietzsche wrote that attending carefully to music can teach us something important about how to love:

> First one has to learn *to hear* a figure and melody at all, to detect and distinguish it, isolate and delimit it as a separate life. Then it requires some exertion and good will to *tolerate* it in spite of its strangeness, to be patient with its appearance and expression, and kindhearted about its oddity. Finally there comes a moment when we are *used* to it, when we wait for it, when we sense that we should miss it if it were missing; and now it continues to compel and enchant us relentlessly until we have become its humble and enraptured lovers who desire nothing better from the world than it and only it. But that happens to us not only in music. That is how we have *learned to love* all things that we now love.[18]

If this is indeed the message of the Golden Record, then perhaps Ann Druyan, the creative director of Sagan's team, is right to call it a love letter from humankind.[19]

The "if" is big. We have no idea whether the unknown recipient of our collective message is capable of returning such feelings. Druyan, who had fallen in love with Sagan during the project, felt compelled to deliver a representative sample of such feelings by recording her brainwaves for the LP while thinking of their blossoming romance. This gesture takes the mixtape, which is always a labor of love, to a whole new level.[20] But perhaps it is not wrong to assume that extraterrestrials might be attuned to love. After all, we cannot but humanize the alien. No matter how hard we try, our imagination is shackled by our experience. This inability is systematic and goes all the way down: as soon as we imagine the alien, we have already missed what is essential to it—the unimaginable. But perhaps missing the point of such an impossible task is precisely what drives our attempts to love the other, particularly through seemingly crackpot ideas such as sending music to communicate with aliens. And who knows? An extraterrestrial might have a Nietzschean response to our music and find its initial strangeness a lesson in love. Music's alienness—its "oddity," as Nietzsche puts it—may arouse a journey from attention to patience to familiarity and then to rapture as the Golden Record is replayed over and over on another planet. Thinking the alien may be a futile contradiction, and yet estrangement may also be the beginning of enchantment. This is music's promise.

Whatever the alien response might be, there is something fitting about sending our "strange" music into outer space. Music is a mystery. We do not know exactly what it is for, which may explain its strangeness. Justifications for music among Earth dwellers have passed the whole gamut from frivolous "auditory cheesecake"[21] to "the most important thing we ever did."[22] But whatever the solution may be to this mystery, music is clearly a deeply human activity found in all cultures and has the capacity to leap over the boundaries of language. The Romantic adage of music as universal language gains new currency in this space-age context; by collating a representative sample of World Music on one disc, NASA symbolically merged the nation-states into "a single global civilization," as if the inhabitants of one planet could speak to another in universal tones.

The astrophysicist Sebastian von Hoerner made the case in 1973 that the specific mathematical properties of the musical scale could be optimized with some small tweaks—what we would call 5-tone, 12-tone, or 31-tone equal temperament—so that it might become usable as a means of interstellar communication.[23] In this way, Earth's global tones become cosmic tones. Von Hoerner proudly claimed later that his short article provided the inspiration for the inclusion of the Golden Record in the Voyager mission.[24] The blockbuster movie *Close Encounters of the Third Kind*, released just a few months after Voyager's launch, even made musical communication with aliens a central moment of its plot, in which the iconic five-note jingle, reproduced in the epigraph to this chapter, somehow turns into an interspecies game of Simon.[25] Not surprisingly, that scene was the only part of the movie that Carl Sagan approved of.

It is conceivable that von Hoerner's mathematical argument may have swayed some scientists to consider music, even though, of course, almost none of the music included in the Golden Record (save perhaps that performed on Glenn Gould's modern Steinway grand piano) possessed the specific musical properties that von Hoerner described in his perfectly even scales. But what mattered more than anything else is that he felt music *might* help with interstellar communication; its mathematical patterns were deemed universal enough to make alien contact feasible. Bach's music is featured three times on the record, precisely because Sagan's team thought its mathematical and contrapuntal structures might "speak" to the alien. And even failing that, the interstellar mixtape still signaled something deeply human to any alien who might care to listen. It is not so much the tones as the genre that seeks a connection. In the 1970s, mixtapes were highly personal, home-spun compilations; they were a form of self-curation. In the form of a cassette, mixtapes were the first portable genre that fashioned music as the sound of personal identity on the go. They defined how humans moved in time and space.

They literally sonified a distinctive way of being in the world. Wound within the cassette was a soundtrack of the self; it was a kind of metamedium that assembled the self from samples of different recordings. These tiny objects were cheaply replicated at home and were given as tokens of identity. The Golden Record, although an LP on the outside, is a mixtape on the inside. It is a token of humanity's self-curated identity, an assemblage of recordings that styles our species. Even its rudimentary mechanism has the feel of a homespun object, lovingly made as a personal soundtrack. As with its iPodic counterparts today, this playlist is to be shared in stereo, with the channels divided between the left and right ears of different species. Let's just hope that Voyager, barreling through the universe like a wireless earbud, finds a friend across the galaxy to share humanity's mixtape (see Figure 3.2).

MUSIC ON THE GOLDEN RECORD: TABLE OF CONTENTS[26]

1. Brandenburg Concerto no. 2 in F Major, BWV 1041: 1. [Allegro]. Composed by Johann Sebastian Bach, performed by Munich Bach Orchestra (featuring Karl-Heinz Schneeberger, violin), conducted by Karl Richter. Recorded in Munich, Germany, January 1967. 4:40.
2. "Ketawang: Puspåwårnå" (Kinds of flowers). Performed by Pura Paku Alaman Palace Orchestra/K.R.T. Wasitodipuro (director) featuring Niken Larasati and Nji Tasti (vocals). Recorded by Robert E. Brown in Yogyakarta, Java, Indonesia, on January 10, 1971. 4:43.
3. "Cengunmé." Performed by Mahi musicians of Benin. Recorded by Charles Duvelle in Savalou, Benin, January 1963. 2:08.
4. Alima Song. Performed by Mbuti of the Ituri Rainforest. Recorded by Colin Turnbull and Francis S. Chapman in the Ituri Rainforest of the Democratic Republic of Congo (Zaire) circa 1951. 0:56.
5. "Barnumbirr" (Morning star) and Moikoi Song. Performed by Tom Djawa (clapsticks), Mudpo (didgeridoo) and Waliparu (vocals). Recorded by Sandra LeBrun Holmes at Milingimbi Mission on Milingimbi Island, off the coast of Arnhem Land, Northern Territory, Australia, 1962. 1:26.
6. "El Cascabel." Composed by Lorenzo Barcelata, performed by Antonio Maciel and Los Aguilillas with Mariachi México de Pepe Villa, conducted by Rafael Carrión. 3:14.
7. "Johnny B. Goode." Written and performed by Chuck Berry (vocals and guitar) with Lafayette Leak (piano), Willie Dixon (bass), and Fred Below (drums). Recorded at Chess Studios, Chicago, IL, on January 6, 1958. 2:38.
8. "Mariuamangi." Performed by Pranis Pandang and Kumbui (mariuamangi) of the Nyaura clan. Recorded by Robert MacLennan in the village of Kandingei, Middle Sepil, Papua New Guinea, on July 23, 1964. 1:20.
9. "Sokaku-Reibo" (Depicting the cranes in their nest). Arranged by Kinko Kurosawa, performed by Goro Yamaguchi (shakuhachi). Recorded in New York City circa 1967. 4:51.
10. Partita for Violin Solo no. 3 in E Major, BWV 1006: 3. "Gavotte en Rondeau." Composed by Johann Sebastian Bach, performed by Arthur Grumiaux. Recorded in Berlin, Germany, November 1960. 2:55.
11. *Die Zauberflöte* (The magic flute), K. 620, Act 2, no. 14: "Der Hölle Rache" (Hell's vengeance). Composed by Wolfgang Amadeus Mozart, performed by the Bavarian State Opera Orchestra and Chorus, featuring Edda Moser (soprano), conducted by Wolfgang Sawallisch. Recorded in Munich, Germany, August 1972. 2:55.
12. "Chakrulo." Performed by Georgian State Merited Ensemble of Folk Song and Dance, featuring Ilia Zakaidze (first tenor) and Rostom Saginashvili (second tenor), directed by Anzor Kavsadze. Recorded at Melodiya Studio in Tbilisi, Georgia. 2:18.

13. Roncandoras and Drums. Performed by musicians from Ancash from recordings collected by José María Arguedas (Casa de la Cultura, Lima) in the Ancash region of Peru, circa 1964. 0:52.
14. "Melancholy Blues." Written by Marty Bloom and Walter Melrose, performed by Louis Armstrong and His Hot Seven. Recorded in Chicago, IL, on May 11, 1927. 3:05.
15. "Muğam." Performed by Kamil Jalilov (balaban). Recorded by Radio Moscow circa 1950. 2:30.
16. *Le sacre du printemps* (The rite of spring). Part 2.6. "Sacrificial Dance (The Chosen One)." Composed and conducted by Igor Stravinsky, performed by Columbia Symphony Orchestra. Recorded at the Ballroom of the St. George Hotel, Brooklyn, NY, on January 6, 1960. 4:35.
17. *The Well-Tempered Clavier*, Book 2: Prelude and Fugue no. 1 in C Major. Composed by Johann Sebastian Bach, performed by Glenn Gould (piano). Recorded at CBS 30th Street Studio in New York City on August 8, 1966. 4:48.
18. Symphony no. 5, op. 67: 1. Allegro con brio. Composed by Ludwig van Beethoven, performed by the Philharmonia Orchestra, conducted Otto Klemperer. Recorded at Kingsway Hall, London, UK, on October 6, 1955. 7:20.
19. "Izlel je Delyo Hagdutin." Performed by Valya Balkanska (vocals), Lazar Kanevski and Stephan Zahmanov (kaba gaidi). Recorded by Martin Koenig and Ethel Raim in Smolyan, Bulgaria, 1968. 4:59.
20. Navajo Night Chant. Performed by Ambrose Roan Horse, Chester Roan, and Tom Roan. Recorded by Willard Rhodes in Pine Springs, Arizona, summer 1942. 0:57.
21. "The Fairie Round" from *Paueans, Galliards, Almains and Other Short Aeirs*. Composed by Anthony Holborne, performed by the Early Music Consort of London, directed by David Munrow. Recorded at Abbey Road Studios, London, UK, September 1973. 1:17.
22. "Naranaratana Kookokoo" (The cry of the megapode bird). Performed by Maniasinimae and Taumaetarau Chieftain Tribe of Oloha and Palasu'u Village community in Small Malaita, Solomon Islands. 1:12.
23. Wedding Song. Performed by a young girl of Huancavelica. Recorded by John and Penny Cohen in Huancavelica, Peru, 1964. 0:38.
24. "Liu Shui" (Flowing streams). Performed by Guan Pinghu (guqin). Recorded before 1977. 7:37.
25. Bhairavi: "Jaat Kahan Ho." Performed by Kesarbai Kerkar (vocals) with harmonium and tabla accompaniment. Recorded in Bombay (Mumbai), India, April 1953. 3:30.
26. "Dark Was the Night, Cold Was the Ground." Written and performed by Blind Willie Johnson (slide guitar, vocals) in Dallas, TX, on December 3, 1927. 3:15.
27. String Quartet no. 13 in B♭ Major, op. 130, 5. Cavatina. Composed by Ludwig van Beethoven, performed by Budapest String Quartet. Recorded at the Library of Congress, Washington, DC, on April 7, 1960. 6:37.

Figure 3.2. In the movie *Guardians of the Galaxy*, Peter Quill (aka Star-Lord) carries a mixtape compiled by his mother with music from the 1970s. It is the one piece of nostalgia that connects his identity to Earth and accompanies him (with the aid of a Walkman) on all his intergalactic adventures. Its function is akin to the Golden Record: a personalized assemblage of music shared as a slice of time, a token of identity, and a gift of love—with added awesomeness.

0010. THE COLLECTION

The selection of music has always been deeply contentious. If a mixtape reveals something profound about our musical identity, and if we are trying to pack everything that is important about our human culture into one record, the stakes are simply astronomical. Every gesture matters. Omissions are as telling as inclusions. It has become something of a parlor game to imagine what else could have been included. Much ink has been spilled over the question of whether the Beatles' "Here Comes the Sun" was originally meant to be included, but failed because of copyright issues.[27] The organizers were clearly conscious of the deeply political nature of their work. They invited UN delegates to contribute greetings—which, in hindsight, turned out to have been a mistake.[28] Some delegates clearly loved the sound of their own voice. The self-indulgent lengths to which some delegates went, in some cases reciting poetry, exceeded by far what the organizers hoped to fit on the record. In the end, the organizers turned the extensive polyglot greetings of our UN representatives into a collective murmur, creating a kind of sound collage, set against the soothing sonic background of whale song, as a solution to this diplomatic minefield that is as elegant and aesthetically pleasing as it is economical. Admittedly, the interspecies mingling will potentially add an element of confusion for extraterrestrial listeners.

The team's second attempt in this regard focused on languages, rather than nations. Carl Sagan asked the language departments at Cornell, his home university, to record greetings in as many languages as they could come up with. The task was open-ended: the native speakers were simply asked to record what they would say if they greeted an extraterrestrial. The results are checkered. Some are perfunctory and general, some are more inviting—extending explicit invitations to visit planet Earth—whereas others issue more cautious greetings from a safe distance. The Indonesian greeting takes no chances by jumping straight ahead to the end of the first encounter: *Selamat malam hadirin sekalian, selamat berpisah dan sampai bertemu lagi dilain waktu*—"Good night, ladies and gentlemen. Goodbye and see you next time."

The strategy that Sagan's team adopted here is helpful in understanding the purpose of the Golden Record. The recordings are gathered as exemplifications of an underlying principle. The record becomes the sonic equivalent of a shadow box, a display case with multiple openings to be filled with specific objects on a given theme that, taken as a whole, represent a narrative by example. It is a collection.

Collections are not just amorphous assemblages of objects. They are a material principle of the structures of organization. Having a complete set of something may be every collector's dream, but in many cases, the organizing principle is so vast that the totality is beyond the scope of an individual collection. We do not need to collect *all* the state quarters that have ever been issued; one coin from each of the fifty states is enough, or even just the most interesting ones. In other words, we group objects together until they can tell a certain story, until the sample size is deemed representative. Its primary concern is classification—each specimen fills one space in a given domain, one opening in the shadow box. The collection of fifty-five languages that the organizers assembled, based on the collective linguistic forces available at Cornell, obviously omits many other languages. It is far from a complete set of all the sixty-five hundred or so spoken languages on Earth (not counting dead languages, of which there are five on the record)—but it is a fairly good representative sample.

And second, collections are radically decontextualized. This is not only true in this most hermetic context, in outer space, but every collection "replaces origin with classification."[29] A selection of representative samples hangs together not because they all come from the same source, but because all the objects are similar in some way. Collections give up any pretense of being grounded outside themselves in favor of an internal coherence. This is why the whale song that was sneakily included in the language collage is likely to throw a wrench in the works: this form of mammalian communication lies outside of the shadow box of human languages, but extraterrestrials would have no useful criteria by which to distinguish it from human languages. Or perhaps they will discover similarities undetected by human linguists and learn to speak "whale."

The same criteria apply to the musical collection. The prompt was to represent the diversity of human music making in its geographic, historical, and cultural depth. Several anecdotal recollections illustrate the considerable effort that went into filling specific slots in the collection. The geopolitical situation required, for instance, that the Soviet Union—and specifically, Russia—be represented musically. The first suggestion, the tenor Nicolai Gedda singing the Russian folk song "Korobeiniki" (Коробейники, which has since begun an afterlife as the theme tune of the classic arcade game *Tetris*), was discarded because the Swedish-born singer was deemed insufficiently Russian. The official Soviet suggestion, the folk song-like film tune "Podmoskovnye vechera" (Подмосковные вечера, 1956), was turned down by the curatorial team because they felt it lacked authenticity.[30] In the end, two pieces of music from other Soviet republics, Azerbaijani balaban (wrongly described as "bagpipes") and Georgian choral music, were included, while Russia itself was somewhat sidelined. Admittedly, the "Sacrificial Dance" from Stravinsky's *Rite of Spring* supplied the collection with some Russian folk elements, albeit in a violently mutilated form.

Historical depth is represented in various ways. The notated history of Western music makes it relatively easy to date and recreate music from past periods; Sagan's team made an effort to represent different centuries, from English Renaissance dances to Stravinsky's *Rite of Spring*. Needless to say, the West does not have exclusive purchase on history. The most ancient music on the record is, by all accounts, the traditional Chinese guqin piece "Liu Shui" 流水 (Flowing streams). This piece is associated with the legendary qin player Bo Ya 伯牙, who may have lived during the Spring and Autumn Period or the Warring States Period (that is, sometime between the seventh and the fourth centuries BCE.) Even if these claims cannot be verified, the earliest notated versions of this piece go back to 1425.[31] And in their efforts to cover a variety of cultures, the curators of the Golden Record also included a range of American popular music, from Blind Willie Johnson's gospel blues "Dark Was the Night" to Chuck Berry's rock 'n' roll hit song "Johnny B. Goode."

It is easy to fault representative collections for all the things that are not included. Any decision to include one piece is always a simultaneous decision to exclude an infinity of other possible choices. To be sure, several flaws have been pointed out. There are age, gender, and geographical disparities. Large swathes of the world, most prominently the entire Middle East, are completely absent from the record. Popular music is exclusively represented by American musicians. And the bias for the Big Bs of classical music—with two pieces by Beethoven and three by Bach—is glaringly apparent.[32] Moreover, the Australian musicians are not recognized, and the piece is misidentified.[33] The full title of Blind Willie Johnson's "Dark Was the Night" (which actually continues in a decidedly unromantic vein:

"Cold Was the Ground") was shortened beyond recognition. A number of pieces were plainly misattributed.[34] All kinds of conscious and unconscious biases abound. But as a collection, the twenty-seven tracks that make up the Golden Record are relatively well chosen and in some ways quite progressive. Choosing Stravinsky's *Rite of Spring* was a notable decision — Ferris explains darkly that this "composition built no bridges between cultures."[35] And the Black trajectory of American popular music is made abundantly clear in the selection at a time when Elvis the King had barely passed away.

There is a certain irony that the most lovingly curated collection of World Music, once confined to the grooves of the Golden Record and on its way to outer space, ceases to be World Music in any meaningful sense and turns into what we might call "Earth Music." This is part and parcel of its status as a collection. Uprooted from its original cultural context, the music has only its internal coherence in the collection to draw on, and it lacks all external reference points. On the Golden Record, it makes no difference whether the music stemmed from an improvisatory tradition or a painstakingly notated one, whether it served in a participatory setting, a ritual procession, a dance party, or a concert hall. The temporal and cultural relations of World Music are erased and reconfigured into the internal relations that make up the collection, and the meaning of these relations must first be renegotiated. While World Music celebrates cultural difference, Earth Music flattens any such distinctions into one homogenous, earthbound culture. On its way into the unknown, the only difference it knows is in comparison with the culture that might pick it up at the other end.

In an unexpected twist, the Golden Record turns all musicking into absolute music writ large. Loosened from its context ("absolute"), detached in the most literal sense from the conditions of its creation, it has nothing left but the musical structures and the inner coherence of the sounds from which it is constituted. As Voyager speeds away from Earth, we must leave behind our human approach to understanding music. Any improvised performance, any interpretation of a score, becomes the thing itself, and not a creative representation of something that exists outside it. Every musical performance included on the Golden Record becomes an autonomous "work," a Platonic instantiation of itself. In outer space, there is nothing outside the Golden Record.

Given such radical autonomy, if an intelligent alien life-form were to intercept NASA's LP and release its frequencies, it would create a new cultural network around the music that will not coincide with the original network on Earth. Meanings do not travel well, and they become meaningless in space, which is why the text messages on Voyager will have no chance of alien communication. If music is to connect the two networks, it will obviate the need for meaningful transmission by turning inward. It will transmit "itself." The means become the end, which makes the means and not the meaning the bridge that links the networks without bringing them into coincidence. Since there is nothing outside the Golden Record, any distinctions must be recreated from within the material that the music provides.[36] We are faced, then, with a fundamental structuralism à la Saussure that operates with nothing more than the two basic categories of identity and difference. And it is precisely at this basic level that music gestures to its intergalactic credentials, for, aptly, this new structuralism corresponds to the binary system of one and zero, which NASA accords universal validity beyond the human sphere. Music blinks: 0101.

It appears that NASA expected the alien recipients to attend "Music 101"—the virtual Introduction to the Fundamentals of Music Analysis course secretly inscribed on the Golden Record. It was originally a "Great Music" course, if the history of the record's curation is anything to go by. When the curators first put together the Golden Record, a number of alternative approaches were suggested. One obvious proposal, though dismissed at an early stage, was a tape recording of Beethoven's Ninth Symphony, with the "Ode to Joy."[37] Another was simply to send the complete works of J. S. Bach into outer space—except "that would be boasting."[38] In the end, though,

Bach is the most amply represented artist on the Golden Record, with three pieces (representing solo instrument, multivoiced instrument, and orchestra), closely followed by Beethoven with two (representing his orchestral and chamber music). The motivation behind this decision, however, had shifted subtly, but decisively: it was no longer an earthly (or, more precisely, a Western) humblebrag, but an attempt to facilitate "'decoding' by extraterrestrial listeners," to give them a basis on which to make meaningful comparisons.[39]

In other words, what NASA was hoping for are alien music analysts. Music 101, as the course number suggests, is a binary exercise in identity and difference—the basis for making "meaningful comparisons." But given that NASA's collection of music is locked in an autonomous world of frequency relations, there is no good reason to assume that extraterrestrials would keep their categories neatly differentiated following human terms. Why should issues such as authorship, chronology, and style trump others such as instrumentation, texture, or speed? Is a slow, intimate movement from a string quartet really recognizably similar to a fast and furious symphonic movement, just because it was written by the same composer? Is a contemplative keyboard prelude and fugue, performed on a modern Steinway grand piano, recognizably similar to the rhythmically driven concerto grosso movement with its timbral kaleidoscope of solo instruments: trumpet, recorder, violin, oboe? It is uncertain if Bach and Beethoven will retain their identity in space.

Music 101

I think there's a bit missing.

Indeed, if we want to reconstruct the mind of NASA's extraterrestrial music analyst, we need to estrange our basic understanding of similarity and difference. On the interstellar turntable, the tables are turned. While World Music, as a celebration of diversity, starts from an assumption of difference, the Earth Music that emerges on the Golden Record begins from an assumption of similarity, of unity—the assumption of a shared culture on Earth that should be compared to and contrasted against whatever exoplanetary culture may find it. From a sufficient distance of several light years, the hocketlike motives of the paired flutes of the Iatmul people in Papua New Guinea may appear similar to the brusquely repetitive blocks of Stravinsky's *Rite of Spring*, and the Scotch snaps of the English Renaissance dances may align with the syncopations of the Alima song of the Mbuti people of the Ituri Rainforest in the Democratic Republic of Congo. These relations make little sense here on Earth—we *know* that they come from completely independent musical traditions. But this kind of cultural knowledge is precisely what falls by the wayside on the Golden Record. The construction of similarity and difference—in short, the construction of identity—must restart from scratch. All that is given is what is stored in the groove of the record.

★ ★ ★

Enveloped in high cultural intentions and sealed with love, the Golden Record is first and foremost a decontextualized storage medium. This shiny round coffin that carries the earthly remains of our music leaves Earth behind as a habitat of meaning and erases the ever-changing nature of music in order to foreground its existence as an object that can be repeated with each playing. Its endless loop circumscribes its own autonomous world. For the alien recipient, this *object*-ive existence is the precondition for studying Earth Music.

In this sense, the golden disc is the music of the spheres in reverse—an autotelic *harmonium mundi* to be deciphered by the universe. It is a microcosmos for alien reception. Aptly, the first music etched on the Golden Record is a computer-generated realization of Johannes Kepler's *Harmonices Mundi* of 1619—a snippet of Laurie Spiegel's composition "Music of the Spheres" opens the section on Earth sounds.[40] Armed with new empirical data, Kepler updated the Pythagorean cosmos by aligning the elliptical motion of the planets with the geometric order of the universe. The six known planets at the time, spinning around the sun, were like the diverse tracks of the Golden Record, their different rotational speeds generating a music between celestial bodies that was no longer monodic, but polyphonic. Kepler's laws of planetary motion issue from a contrapuntal universe, with Earth, Mars, Venus, Jupiter, Saturn, and Mercury operating in different vocal ranges and modalities, as shown in Figure 3.3. Similarly, the Earth Music on the Golden Record may appear, at first, to be a fixed monadic entity endlessly rotating in its own orbit, but it is in calculating the difference between the tracks that an alien recipient would build up a polyphonic soundscape of our planet. This LP is a *mixed* tape, after all, and each track offers a different range and modality from which to reassemble a world, which, like the six planets of Kepler's cosmos, can only be partial in its universality.

Figure 3.3. By measuring the angular velocity of the six known planets (and the moon), Kepler surmised that the universe operates as an interplanetary choir. Venus, an alto along with Earth, manages to rise only a languorous quarter tone that goes unnotated. Mercury, the only soprano, is obviously something of a diva. From Johannes Kepler, *Harmonices Mundi* (1619).

NASA's *harmonium mundi* shares another trait with Kepler's *Harmonices Mundi*. The astronomer's calculations, unlike Spiegel's realization, were never designed to be heard: they were inscriptions for the human intellect, data measured by those made in the image of God to capture an imprint of the mind of God. NASA's golden disc is also silent data, an imprint made in the image of humanity. It does not have to be heard, for the recorded music is not transmitted as sound, but as inscriptions of sound waves. There is nothing other than the notation of frequencies on the disc. Music's objective existence is literally an object for analysis. So in comparing the different tracks, the alien analyst sitting in Music 101 will not be studying music as we know it. What is assumed as "normal" in music theory vanishes in space. To get a better sense of what alien music analysis might look like, we need to take into account its mediality. We need to reverse engineer the Golden Record.

CHAPTER FOUR

Transmission

E.T. phone home.
—*E.T. the Extra-Terrestrial*

Vagaries in communication are not only a problem in interstellar space. When the sociologist and system theorist Niklas Luhmann coined the sobering axiom "communication is improbable,"[1] he wasn't casting his eyes toward the heavens. He was merely thinking about the vicissitudes of communication between human senders and receivers. But clearly, Luhmann's axiom governs the Voyager project. In beaming our musical message into the starry sky, we have absolutely no idea where the spacecraft is going, when we would make contact, or with whom we are communicating. Communication in this scenario is *infinitesimally* improbable. NASA is not even certain that its message is clear. Their earthbound documentation of the Golden Record was published under the telling title *Murmurs of Earth*, which seems to suggest that Sagan and his team were all Luhmannians: "murmurs" are inchoate mumblings that assume that "communication is improbable."

But it is not all bleak. Luhmann suggests that the chances of communication from sender to recipient, from *ego* to *alter*, can be exponentially improved thanks to the work of media. Media jump to the rescue in the most literal way imaginable, by coming *between* the two sides and channeling the communication from beginning to end. In the case of the Voyager mission, with the human collective *ego* reaching out to an extraterrestrial *alter*, the situation is a bit more complicated than usual, but not infinitely so.

0001. ALIEN MEDIATION

Ostensibly, Sagan's model of communication, more loosely conceived than Luhmann's, follows a commonsensical telephone model: a message is sent, transmitted, and received. However, given the vast distances to cover and the relatively slow speed a human-made spacecraft travels, a "baby civilization like ours" is more likely to be at the receiving end of the telephone message.[2] After all, most of the universe has had an evolutionary head start on alien communication, because our blue planet is an estimated 4.54 billion years young, compared with the approximately 13.4 billion years of the Milky Way. From the perspective of the big other, our little space probe is crawling across the galaxy like a toy. Meanwhile, awaiting a call from beyond, the observatories at Arecibo, Green Bank, and elsewhere are set up as gigantic ears to listen constantly for any incoming radio waves from more advanced civilizations. Transmission by radio wave has the advantage of traveling at the speed of light, but this form of messaging has to be temporally precise — if no one's recording when the radio waves reach our planet, we will miss the call.

Sagan's ambition to punch above the baby weight of our civilization and to be a sender is a more specific and precarious form of communication. Although Earth is constantly emitting radio waves, Sagan clearly wanted to send something intentional and concrete, directly addressed to the unknown alien to express something about us. As a sender with a physical payload, the Voyager model of communication is based on spatial precision: it is not impossible to score a hit, but given the vastness of space, the tininess of the space probe, and the myriad of uninhabited obstacles in the way, a miss is highly likely. As Sagan concedes, the two Voyager spacecraft are wild shots in the dark.[3]

But even if there is a hit, how would we engage in communication? We have no idea how to interface with the receiver. Unless we share a medium, transmission will fail. Sagan believed there is at least one common "language" between our species and any extraterrestrial civilization that Voyager descends on: science.[4] Science is the interface of communication. Sagan's boundless enthusiasm for a transcendent science is both pragmatic and an ideological hallmark of his work, and there are good reasons for promoting science as the interface in our quest for interstellar communication. Science is grounded in numbers. Numbers are its lingua franca, specifically, the simplicity of binary numbers, with their repetition of identity and difference. Counting in binary patterns, then, is the key in a renewed *mathesis universalis* of alien contact.

PREMISE ONE

Numbers are the medium of communication with other intelligent life-forms across the universe. All numbers can be represented as binaries (1/0). This system of identity and difference is the basis for repetition.

But numbers do not count by themselves. They have to be made present, as if counting out loud. We may share a "language" of numbers with the alien, but if we do not interface via a physical medium to transduce the data, there is no contact; the numbers will remain abstract and immobile, analogous to eternal integers of a Pythagorean cosmos. For transmission to occur, numbers need to materialize and move in time. They need to vibrate. The search for extraterrestrial life has suggested a few constants that life on Earth might have in common with life on other planets. Of particular interest are planets with water, the precondition for all life as we know it, although ammonia, sulfuric acid, and methane are also possible alternatives. Sentient beings elsewhere in the universe are almost certainly carbon based—or maybe, just maybe, silicon based.[5] And they will likely exist in a fluid medium, such as a liquid sea or a gaseous atmosphere. More pertinent to our exomusicological mission, sensory perception on other planets is expected to be significantly reliant on vibrations, oscillations, and waves, as it is among sentient beings on Earth. These sensory constants are the media for the message. They channel the transmission of numbers, which SETI, the group of scientists specifically interested in the search for extraterrestrial intelligence, expects to be the binary system of on-and-off impulses universally understood by all forms of intelligent life.

These constants set some basic parameters for alien communication. There are three factors at play here, somewhat akin to the telephone model. For E.T. to phone home, the system would require:

1. *Sender:* vibrations and oscillations
2. *Transmitter:* the fluid medium through which these waves are propagated
3. *Receiver:* the physiological receptors of carbon-based intelligences

All three constants rely on counting and repetition. They send, transmit, and receive numbers. Hence, the music theory outlined in our blueprint is based on frequency, which, in its simplest formula, is counting repetition. It blinks on and off as a binary system. And aptly, NASA has sent a frequency machine as the transmitting device to cajole the alien to count aloud with us.

0010. MUSIC MACHINE

Yet why is music transmitted as a machine? Given that the transmitter in our telephone model is a "fluid medium through which these waves are propagated," a machine seems too detached, discrete, and solid a medium to transmit the vibrations.

The reason for the machine is time and distance. Counting takes time. When music counts aloud, it is an event; music happens, then disappears irrecoverably down the temporal vortex into the past. An event cannot time travel. But as the media philosopher Sybille Krämer points out, all recording media spatialize temporalities.[6] Time solidifies, and data move into storage, where the numbers no longer count, but are suspended in space. In this state, it can travel in time and cross vast distances.

Fixed as data within a machine, the numbers operate differently: they become analyzable. What would have evaporated in time can now move in all directions in the same way that a spoken sentence is irretrievable in time, but when written on a page can be edited, reordered, or erased. For an alien, the machine is the site for music analysis. The frequency machine on board Voyager, then, is not an addition to music, but integral to its machine being. This being is highly complex, because media are seldom, if ever, pure; they function as various machines attached and detached in different configurations, storing, transforming, and transferring data through other media in which they are submerged. In this sense, the entire spacecraft represents an assemblage of machines tasked with the transmission of Earth Music. It comes between the two sides, channeling communication from sender to receiver. The task of a media archaeology is to sift through the machinery to excavate the music suspended in transmission.

PREMISE TWO

Exomusicology accounts for music's repetitive cycles in terms of frequency. Frequencies are a form of counting and comparison—a numbering—that can be stored in a machine as data and played back as an event.

So what are the technical components, the materialities of communication, and the media machines, in this interstellar context? There is the space probe functioning as carrier and the glittering LP, of course, encoded with data. But the Golden Record has stolen the limelight from another piece of equipment, a mundane machine that might be the real hero here: the cartridge and stylus. This unit is tucked modestly behind the disc on the outside of the Voyager spacecraft. It functions as a protogramophone stripped down to its essential components, without which the Golden Record would just be a silent copper disc, gilded to avoid corrosion, with grooves etched into its surfaces to store the raw data of human culture. For all that the potential aliens know, the record might just be a round, shiny ornament, an unusually crunchy cookie, or the perfect projectile for interstellar Frisbee.

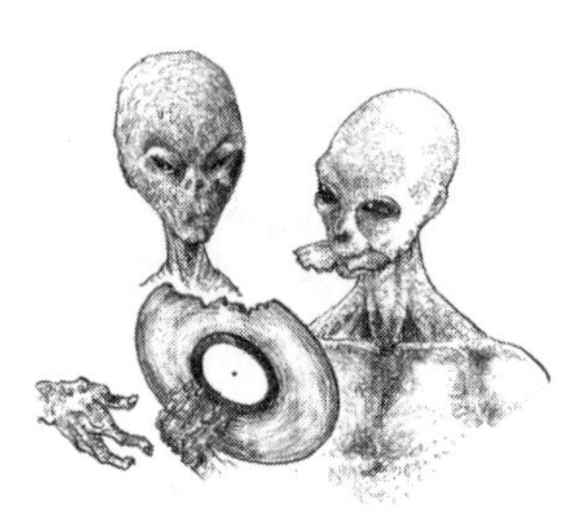

Mmmm . . . Marty, tell your musicologist friend that I worked out what this thing is.

The stylus combats Luhmannian communicative entropy by reconverting the jittery circular line of the gramophone groove into pressure changes in whatever fluid medium the alien civilization resides. It channels the message on the receiving end and acts as a gateway through which frequencies are made perceptible as a basis of communication. NASA's technology demands a certain (con)tactility to complete the handshake between the humans with the record at one end and the alien with the stylus on the other. It brings us into resonance. The stylus is a prosthetic extension of transmission, a medium stretched beyond the human limits of time and space to convey the tiniest tingle — a mere murmur — of Earth-shattering significance. It is not the vibration per se that speaks volumes: it is touch. Contact is the underlying premise of vibrations, which means that vibrations do not make music until they make contact. Indeed, the stylus is the tool that closes the loop, reanimating the data suspended in golden storage to a strange receptor. This recursive moment is a metaloop that folds the alien other into our loop. The tiniest point in

NASA's design, then, is the opening of a vast new network of communication. This is the exoplanetary contact point for an exoplanetary counterpoint.

As such, music does not exist everywhere in the universe, suspended in timeless abstraction, as is often assumed by philosophies of cosmic harmony. Rather, music *can* exist anywhere, insofar as it is stored in a medium, but this "anywhere" can be accessed only as a point, that is, as a coordinate in space-time as a particular relation of contact. After all, this is the underlying mechanics of musical media. Music technology transports music in time and relocates music in space; the point of repeat delivers Voyager's payload.

The contact point, then, is where music is delivered as an *event*—as a happening that takes time and takes place. It switches communication on. Whereas frequencies suspended in storage can exist anywhere and anytime, an event has to be *this* time and *this* place and no other. It is a specific vibration, a frequency realized in space-time as a point. Because such communication has to be temporally and spatially precise, making alien contact is a tricky maneuver, or, to recall Sagan's comment, "a shot in the dark." Transmission is easy; contact is hard.

PREMISE THREE

Vibrations are the concrete expression of numbers. They are vehicles of frequencies at the point of contact. Frequencies can be stored as data. Vibrations cannot be stored, because they take time; they are the stuff of events.

Even if an alien intercepted the space probe and located the stylus, getting the gramophone to function would still be like finding a needle in NASA's haystack of assumptions. The needle may make contact, but the disc has to rotate precisely to deliver the information, otherwise contact would still make communication improbable.[7] To this end, NASA provides an IKEA-style assembly sheet, with wordless instructions. (These instructions, engraved on the protective aluminum cover, are shown in Figure 1.3. NASA also produced a diagram with verbal explanations for humans, shown here as Figure 4.1.) Deciphering its symbolic language requires a high degree of intelligence and a goodly amount of detective work, since NASA builds on its assumption that potential recipients of the Golden Record would possess detailed knowledge of physics and chemistry.

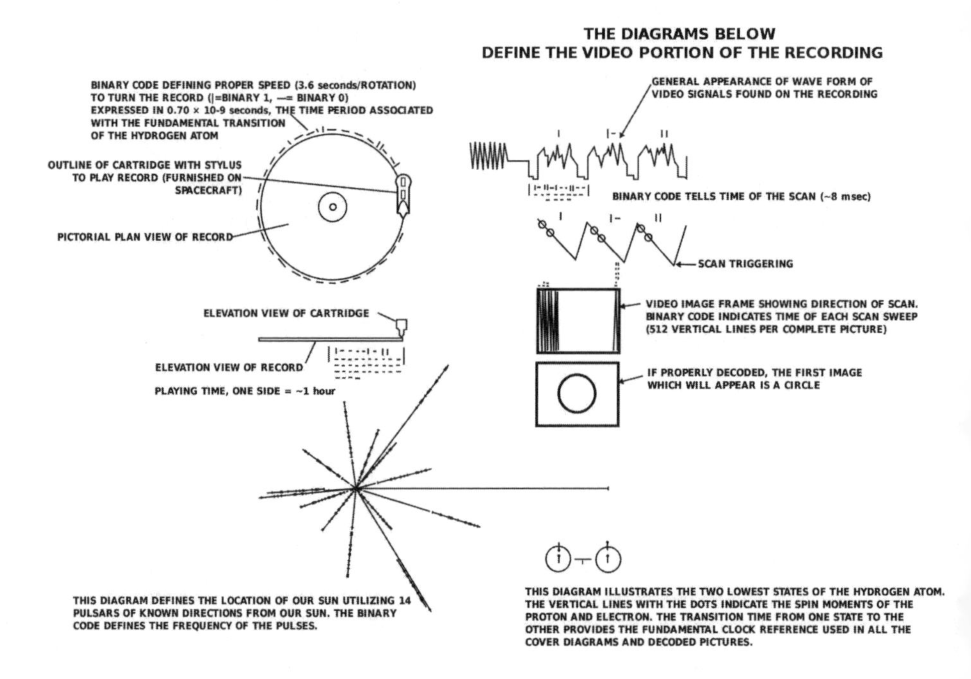

Figure 4.1. NASA's solution to the riddle it posed to extraterrestrials who might find the Golden Record (image: NASA/JPL).

They would need to understand, for one thing, in the bottom right corner of Figure 4.1, that the two circles, each marked with a dash in the center and another dash on the perimeter and linked by a short line, represent a hydrogen bond with a single electron circling a single proton in each half of the molecule. If you look very closely, you can see that the central lines carry a dot at the top of the line, and the outer lines show a dot at the bottom (on the left) and top (on the right). Our intelligent life-form must then recognize this representation as a hyperfine (or spin-flip) transition of hydrogen. In other words, they must know the twenty-one-centimeter line.

This concept might not be a household term, but it is a staple of modern physics: every time an electron whizzing around the proton of a hydrogen molecule changes its spin, that is, when it "flips" from parallel to antiparallel or vice versa, it emits a wave of electromagnetic radiation. This phenomenon can be observed over great distances and is consistent across the galaxy. As such, the hyperfine transition is the closest thing we have to an exoplanetary yardstick. More specifically, our alien scientists have to recognize the time this emission takes: this extremely short duration, 0.7×10^{-9} seconds, is designated here to function as the unit of time for the fundamental interstellar rpm.[8] To return to the image, the tiny line descending from the covalent bond between both hydrogen atoms indicates a "one"—meaning, in this case, the "unit."

This is the key to understanding everything else: all the numbers stated on the instruction sheet, given in binary code, marked as dashes and lines, are multiples of this basic temporal unit. If the aliens do not figure out this constant, they will not be able to understand anything at all. NASA's assumptions may sound outlandish—after all, it is necessary to have an advanced degree in physics to make any headway with this puzzle at all. But it is not unreasonable to presume, if an interstellar message were intercepted by another space-faring species, that an alien society would pour considerable resources into its decoding. One imagines that any team of heroic intergalactic musicologists deciphering the Golden Record would be ably assisted by a team of menial rocket scientists.

On Earth, care must be taken when following the instructions, since the interstellar rpm for the Golden Record does not follow human convention. This is no ordinary disc, and much knowledge had to be compressed into its tiny dimensions if the information was not going to spill over into a double album. To maximize the playing time, the recording was modified: the recording speed was slowed down to 16-2/3, that is, half the standard speed of the common LP format with its 33-1/3 rpm rotation. In this way, the recording time was doubled, while the loss of fidelity was deemed within a bearable range. Bearable, that is, according to human auditory standards.

Even back in 1977, gramophone recordings were already outmoded technology—a medium presented by Edison a full hundred years earlier, in 1877. The state-of-the-art recording technology was magnetic tape. Why, then, did NASA revert to this older device? Because the record was more durable and simpler to operate than magnetic tape. This is why the early idea of placing a cassette tape with Beethoven's Ninth Symphony behind a metal plaque on the Voyager spacecraft never made it past the drawing board.[9] Especially when strung to the outside of a spaceship, exposed to radiation and extreme temperatures, the magnetized tape would have decayed. The resilient materiality of the disc engraving won the day. The decision was based on a kind of reverse archaeology. What would the alien find in the distant future—data, scrambled like spaghetti, fit for a dog's breakfast, or a shiny record that, like an ancient clay tablet, has a message permanently etched in its hard surface to convey its master's voice?

Obviously, in launching a time capsule into space, NASA's concern with the Golden Record is less about high fidelity than about high sustainability. The object has to last as long as possible. After all, only what survives can be remembered. Archaeology is principally about *things* that last, even if knowledge is not passed on. In inviting aliens to be music analysts, the makers of the Golden Record also invited them first to be media archaeologists, because music does not survive as intellectual currency, but as a material object.[10] Our network of knowledge around the music on the Golden Record will dematerialize in time, but once the media are un-*Earth*ed by the aliens, they will make new networks around the object. This is why today's "virtual" knowledge symbolized by "the cloud" is not how we will be remembered after we are extinct. There will be no trace of humanity in the cloud in a hundred thousand years' time. Clouds evaporate. NASA's anniversary project, to send a collective tweet into outer space on the fortieth anniversary of the Golden Record in 2017 under the hashtag #MessageToVoyager, is of no further relevance precisely because there is no *thing* to show for it in the future.

To summarize: in exomusicology, music is first a thing, a technology, a number machine. Despite the space-age aura, these things are not on the cutting edge of scientific progress, let alone futuristic marvels. They are makeshift and rudimentary, designed to survive time. They are almost already the primitive relics that they will certainly become in the future. Thus, in exomusicology, music analysis is as much a question of media archaeology as it is one of music theory. An archaeology reveals the obdurate, resistant, material processes as it disassembles music into its component parts. It demonstrates how music, as an assemblage of machines, assumes various forms and undergoes numerous transformations through a series of interfaces and operations. It analyzes how music moves in time and space as a binary structure, not just internally as a system of identity and difference, but externally as it morphs from frequency to vibration, from data storage to event, from object to instance. To imagine an alien reception of music, then, is to analyze its sound technology in order to reconstruct an alien epistemology of "listening."

PREMISE FOUR

Given its timescale, exomusicology transcends methodologies that presume the temporalities of human audition and underestimate the material afterlife of musical media. Since media archaeology builds on the remnants of past technology by reverse engineering, exomusicology thinks back from the future. From this perspective, *we* are currently the past and exomusicology reverse engineers into the future.

OOII. PHONOGRAPHY

Alien communication has been propelled into the future in the media archaeology of the Voyager mission. But even if the aliens reverse engineered the frequency machine, there is one gap that NASA cannot bridge: What does alien listening sound like? The only answer we know to this question is that we don't know. The possibilities are endless. The alien who intercepts the record may have a brain the size of a planet or thirteen ears the size of a peanut distributed according to the Fibonacci series under its armpits — and who knows what that means for their listening habits and musical tastes. But this doesn't mean we have to throw our hands up in despair. What we can do is to extract meaning from the elements that NASA has given us, examining its material traces to get a closer sense of what listening in this extreme situation might mean. Put differently, what knowledge does NASA's technology offer in the way it channels communication between species?

To start, imagine the work of the Golden Record as a series of interfaces that pass on information across shared boundaries, each of which requires a transformation as it maps data in different ways. Let's plot this out, as in Figure 4.2. If we begin, as we typically do, with musicians producing sounds, which are then recorded and engraved on the LP, we move along the left side of the graph, covering the essential stations of the recording process.[11] Next, the reproduction cycle: here on Earth, this cycle typically covers reading, sounding, and listening, corresponding to our earlier telephone model of sender, transmission, and receiver. The stylus takes us across the boundaries of the recording cycle and initiates reproduction, but what comes afterward remains unspecified here. The interface with alien "ears" remains a blank. This missing link in the chain from gramophone groove to musical listening can be seen as a metaphor for the dazzling range of possibilities that alien listening entails. The furthest interface Voyager can determine is the stylus that allows the grooves of the gramophone record to be turned into vibrations.[12] The stylus is the final frontier.

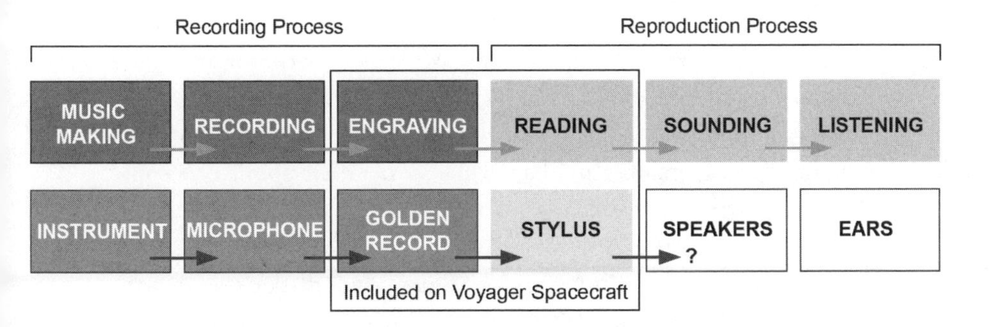

Figure 4.2. Diagram of media and interfaces spanning the recording and reproduction processes of the Golden Record. Since we are dealing with potential alien listeners, the reproduction process was left incomplete by the makers of the Golden Record.

But when the stylus touches the record, the information that it traces from its grooves will be decidedly odd, if not strangely garbled. The final frontier is an interfacial challenge because music is not the only medium that is being transmitted into outer space by NASA. As we have noted, there are also verbose messages by UN delegates, brief greetings in fifty-five languages from random humans, a collage of sounds of the world, and 115 images of our planet. From the perspective of streamlining communication, this babel of expressive forms will likely be a source of confusion. From a media-theoretical perspective, however, this diversity etched on a single disc is a technological masterstroke that illuminates the process of interstellar transmission. It is not so much the contents as the media of communication that reveal the components in an alien epistemology of listening. NASA's LP, then, is a media archaeological site, richly sedimented with knowledge. So having examined the stylus and the instruction for playback, it is time for us to land on the surface of the Golden Record itself and to mine its pits and dig into its grooves.

Gramophone records are a peculiar medium. For the media theorist Friedrich Kittler, they constitute a form of writing.[13] However, their "scripts" do not turn letters into words or dots into melodies; rather, a record leaves a trace of the sound wave on its material surface, faithfully transcribing the pressure changes of the air molecules that we register as sound. Like writing, recordings enable the three essential functions that characterize all technical media: recording, storing, and transmitting data. What is peculiar about this sonic writing is that it is legible to the touch of the stylus.

The gramophone forms one-third of what Kittler calls "discourse networks," circa 1900. Simplifying somewhat, a discourse network constitutes the media structures by which we make sense of the world. For Kittler, the situation around 1900 was determined by a division of writing-down systems into medium-specific technologies: gramophone, film, and typewriter. Each of the three technologies specializes in storing only one expressive form — sound, image, word — and each is specialized in one sensory modality, one way of making sense of the world: listening, looking, reading. Each new technology was extremely efficient in its own domain, far surpassing what had been possible with paper and ink, the media that had dominated the previous discourse network in all three domains.

To understand what phonography changed in the field of music, consider Western musical notation. This form of writing is very efficient at some tasks and quite rudimentary at others. Naturally, most of the parameters essential to Western music are the ones that are easily notated in this diastematic script — primarily, pitch and rhythm. If we try to notate other musical traditions in Western notation, we scramble to get the nuances right. Figure 4.3 shows the Chinese tablature notation for the guqin, for the traditional piece "Liu Shui" 流水 (Flowing streams). Qin tablature is very specific about attack, articulation, and tuning and very loose about rhythmic structure in a way that cannot easily be translated into the grid of Western diastematic notation: our standard notation offers the wrong interface.[14] It is not impossible to notate these sounds, but these dimensions lack the specificity accorded to pitch and metrically organized rhythms, as the two divergent transcriptions of "Liu Shui" show.

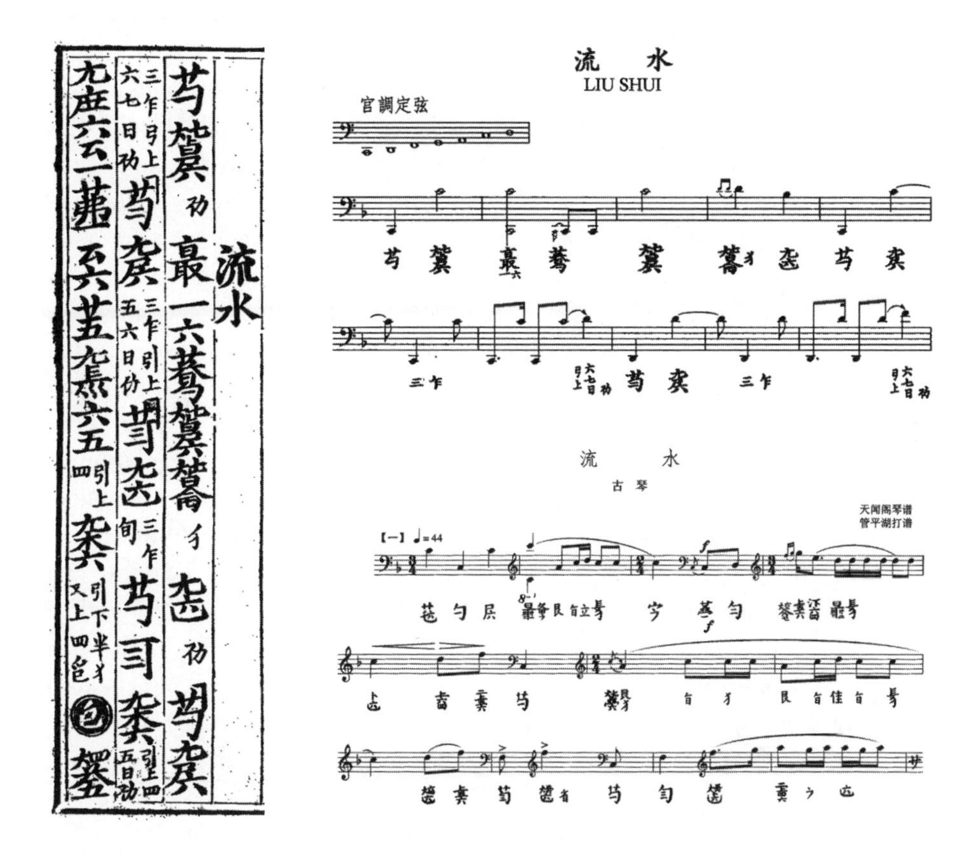

Figure 4.3. Emic and etic notations of the opening of the traditional guqin piece "Liu Shui" (Flowing streams). *Left*: Traditional qin tablature, here a Ming source dating from 1425, uses Chinese characters to notate music. *Quxian shen qi mi pu: San juan (Zhu Quan ji)* 臞仙神奇秘譜: 三卷 (朱權輯). Facsimile reprint (Beijing: Guo jia tu shu guan chu ban she, 2013. 北京市: 國家圖書館出版社, 2013). *Right*: Two modern transcriptions of Liu Shui. From *Shen qi mi pu yue quan*, ed. Wu Wen'guang (Shanghai: Shanghai Music Publishing House, 2008), and *Shen qi mi pu: The Transcriptions/*[*Qu Xian*]; *Silk String Zither Music*, ed. John Thompson (Hong Kong: Toadall Sound, 2001). Thanks to Lingwei Qiu.

The whole project of notation is carried by an ekphrastic hope for translatability between different media, the blind trust that there can be a one-to-one correspondence between them, with the difference crossed by a leap of faith.[15] Such blind trust is not always warranted: there is a certain normative notational logic that goes hand in hand with the music that is supposed to be encoded, carrying within it certain tacit assumptions about the surrounding musical tradition. It is perhaps most useful to think of written notation in the manner of a filter, or better yet, as a kind of grid or cookie cutter: only certain elements can pass through it, and the material is reshaped to fit the mold.

For a gramophone recording, it is not necessary to understand the notational system, that is, the underlying logic on which the encoded music is based. The stylus does not need to have recourse to the abstract and idealized symbols that are instructions outlining the necessary steps for the reproduction of sound. The trace of the needle is an impression of the sound wave.[16] Its notation is simultaneously its performance. But as a consequence, the gramophone recording does not discriminate between music and nonmusic. If, during a recording session, a chair falls over, or a piano lid slams, or a musician starts coughing, all these noises are dispassionately etched into the record groove, whether they are supposed to be there or not. In gramophonic "script," there is no right or wrong. What exists is only sound. There is nothing other than frequencies. The gramophone record, then, because it has no filter function, claims Kittler, permits anything and everything to happen; there is no "cookie cutter" through which sounds must pass before they can be stored.

In other words, the gramophonic surface is ontologically flat; and the flip side of its indiscriminate flatness is its universal script. Hence, on the surface of the Golden Record it is possible to notate specimens of diverse musical traditions in the same space, coexisting side by side as different tracks, where other notational systems would have necessarily failed. The Earth Music that the Golden Record captures is possible only due to the permissiveness of the recording technology.

This is because what the record notates so dispassionately is not the curated content, but time. Or more precisely, it notates a slice of time. And this slice has a peculiar temporal quality: the recording can be infinitely repeated. It keeps time turning in upon itself, running backward to stand still into the future, as if this slice of time were a permanent fixture. It is this recursive loop that limits the boundaries of its flat ontology, compressing the content within its grooves like a flat-ontological microcosm.

Of course, NASA has not pressed only music onto the disc; our entire world is being delivered through the alien letter box as a kind of flat Earth. There are all kinds of incompatible things scattered indiscriminately on its shiny surface. Whereas for Kittler, writing-down systems circa 1900 meant medium-specific technologies storing only one expressive form — sound, image, word — the Golden Record attempts an all-embracing reunion that mends the rifts of modern media. Sound, image, and word coexist in its grooves. Because this is a flat-ontological disc, this reunion is not some Wagnerian *Gesamtkunstwerk* in which the arts collaborate in synthesis: on the Golden Record, sound, word, and image are simply an assemblage of things. In this way, the Golden Record catapults Kittlerian media theory into the fourth dimension in the context of interstellar communication: in the grooves of the Golden Record, the trinity of expressive forms — sound, word, and image — don't have a separate existence; they all become sound.

And to match the boundless medial ambition of the Voyager record, what the Golden Record carries into outer space is not simply a snapshot of human culture; the assemblage of things on the Golden Record contains *all* the human culture that extraterrestrials are likely to encounter.[17] Any alien musicologist that tries to decipher the contents of the Golden Record has only the record groove to work on. It circumscribes the world within a wiggly line, from which the entire culture of our planet must be reconstructed.

0100. WORDS, SOUNDS, IMAGES

Given an ontological surface that flattens diverse media into the same patterns of data, we should analyze the difference in the three domains assembled on the record — words, images, and sounds — to reconstruct an alien encounter with the data.

Words On the Golden Record, words exist either as recorded sounds, most clearly articulated in the parade of languages greeting the potential alien, or as text, most notably in the address by the American president, Jimmy Carter, in the form of a typescript.[18] For all their usefulness on Earth, words are the most useless form of expression in this boundless context. One of the most charming of the greetings, in Southern Min dialect spoken in Fujian province in southeast China, illustrates the problem well: "Friends of space, how are you all? Have you eaten yet? Come visit us if you have time." This all sounds very nice, but what if the aliens haven't yet eaten? What if they are hungry? Might they come and eat us, given such a polite invitation?[19] Indeed, would not eating us be a faux pas? Words, because they refer to things and concepts outside of themselves, lose meaning when they are placed in a decontextualized, self-referential context. The Southern Min dialect, for example, depends on a repository of cultural assumptions that are left outside of what can be contained within the Golden Record: what literally means "Have you eaten your fill?" [你吃了吗] has come to signify "Hello, how are you?" This kind of concern makes intuitive sense to Earth dwellers, with their need for regular food intake, but the required transference, the "carrying across" that all metaphors demand, might not mean much to extraterrestrial cultures, even if they could cut through the linguistic complexities on the sole basis of this short sample.

Clearly, the words on the Golden Record should not be taken literally, since the extraterrestrials cannot decipher the alphabet or connect arbitrary signs to specific meanings from the microcosm of human culture sent by NASA. The verbal components of the message are by far the most symbolic aspect of the Golden Record. This is perfectly summarized by the stealthy, hand-written dedication in the run-out groove: "To the makers of music — all worlds, all times." This inscription, made in obvious breach of the directive to avoid verbal description around the spacecraft, is a performative contradiction: the all-embracing cosmic greeting will be intelligible only to a narrow band of English-speaking humans.

With their extreme limitations, words do not occupy much of a separate domain in the Golden Record, spilling rather into the sonic and visual realms. Hence, the typed presidential address is encoded on the record as an image. Another intriguing image, reproduced in Figure 4.4, is a page from the score of Beethoven's Cavatina from the B♭ Major String Quartet, op. 130, complete with a picture of a diminutive violin, the lead instrument that "sings" in the movement. This page of notation is the musical equivalent of President Carter's speech in typescript. Sagan's team, by placing the score of the Cavatina and its recording last on their respective lists, clearly hoped that the aliens might connect the dots between sound and image, and to nudge the extraterrestrials toward this ekphrastic leap, they further included a snippet of the music after the image. And perhaps to ensure that an alien would not mistake the "sounding dots" as the product of one little violin, the preceding image in the collection is a photograph of the Quartetto Italiano. But how would the alien see the images encoded as sound?

Figure 4.4. The beginning of the score of Beethoven's Cavatina, with an image of a violin (not to scale) (image: National Astronomy and Ionosphere Center, Cornell University; artist: Jon Lomberg).

Images Although sound is usually expected when a record spins on a turntable, there is no necessity for sounds to be the ultimate purpose of a record. Data are just data. Sequences of zeroes and ones can encode images, as well as sounds. Both sides of the Golden Record appear similar, but the B-side contains a rather different kind of "music" than the A-side. One part of its groove is reserved for encoding a range of pixilated images. This means that the analog record is encoding digital data on the B-side, and the stylus, as it spirals across the disc, will unreel the visual information in time, pixel by pixel and line by line, as an event — presuming, of course, that the aliens have already invented a viewing device. A test image is given among the instructions on the aluminum cover, depicted above in Figure 4.1 on the right-hand side. This simple circle within a rectangular frame serves as a calibration page. Another image, engraved just above the framed circle in the diagram, indicates that the data stream must be mapped across the matrix from top to bottom and from left to right. Intelligent extraterrestrials will figure out that each "blip" of the groove represents a dot in a rectangular matrix of unspecific dimensions, and by knowing that the resulting image is a circle, they should be able to determine the size of the matrix. To get technical for a moment: the correct image emerges in a 19 × 29 matrix, which is close to a 2:3 ratio, but using prime numbers for the horizontal and vertical axes. The prime numbers ensure that any other mapping will result in gobbledygook. The overall number of pixels per image is 551 (= 19 × 29). Color images are three times longer, mapping magenta, yellow, and blue in separate passes across the matrix. The beauty of the circle is that it has sufficient symmetry to be recognizable as a deliberate object, but each line, both horizontal and vertical, is distinct from what comes before and after. Errors in mapping and translation will be immediately apparent. If our intelligent extraterrestrial recognizes that the first block of sound on the B-side corresponds to the circular image in the instructions, the alien will have cracked the encryption code. Something resembling Figure 4.5 should appear.[20] Any of the subsequent images, which are much more complex, follow the same pattern. This code is in effect the bandpass filter, the cookie-cutter grid, through which the data are sent.

Figure 4.5. Visualization by Ron Berry of the sonified circle in the Golden Record.

Sounds What is critical here is that the visual data encrypted in a sound wave and etched into a gramophone record are designed to be reconstructed as an image, *but can also be sounded*. Digital data, zeroes and ones, do not have any intrinsic meaning, but can be turned into different expressive forms—sounds or images. So the vibrations that our limited human perspective encourages us to call sound waves are not correlated to a sensory modality that is intrinsically preferable over another. We can *hear* the photo. It sounds like a low machine buzz at 30 Hz, recognizable as a pitch somewhere in the middle between $B\flat_1$ and B_1. If the pitch fluctuates, as it does in an occasional flutter, it is at the octave. Every eight seconds, the continuous hum is punctuated by two test beeps (which demarcate the beginning and end of the pixel data), five octaves higher, around B_5. As a consequence, Beethoven's B♭ Major Quartet, op. 130 is no longer quite in B♭. Whatever kind of music this might be, it is certainly not "great music"—but then again, by what intergalactic standards should we evaluate Earth Music anyway? Aliens might find this hum the perfect buzz for a trance state and never get to see what is ultimately an unintelligible image of diastematic notation with a shrunken violin.

So, strangely, there are actually two representations of Beethoven's Cavatina on board Voyager: one is a recording of a string quartet, the other is a recording of an image of sheet music. Both can be heard through our human ears, if we so choose, but they sound very different. So we should adapt our earlier chart of interfaces, in Figure 4.6, to reflect the realm of possibilities opened up by data conveyed on the Golden Record. Where we could follow the path from recording to reproduction by means of the stylus, the situation has become a whole lot more complicated now. If we can sonify visual and auditory artifacts, we have two different routes of passage that both converge in the gramophone and the stylus, but the outcome becomes increasingly less specific as far as the sensory domain is concerned. In specifying ears and eyes in the far right box of the diagram, we remain within human sensory modalities, but it would perhaps be best to refer to this as "perception" most broadly conceived.

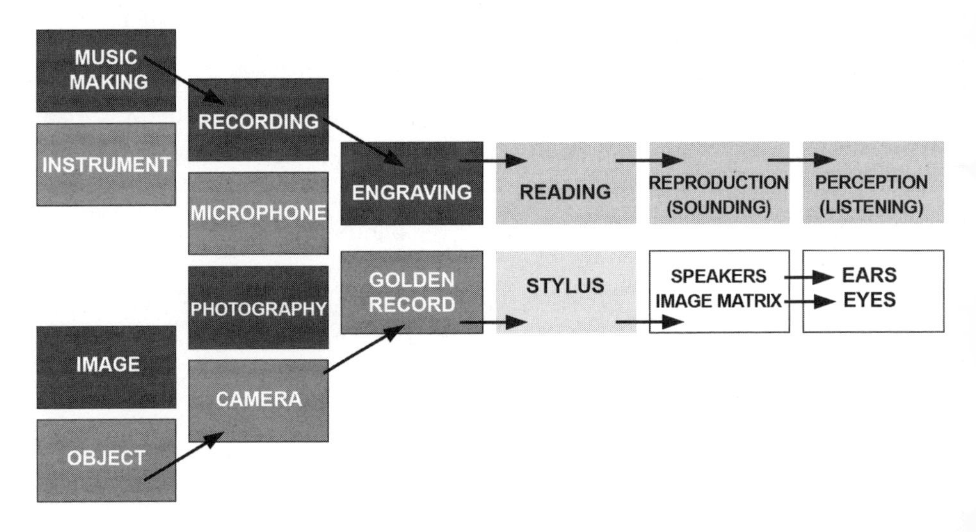

Figure 4.6. The updated diagram of recording and reproduction cycles of the Golden Record. Words lose their distinctive identity on the record and are subsumed under sound (or, in the case of President Carter's typed message, images).

We will get furthest here, in this most interstellar of contexts, if we stop thinking about the sounds as music and start thinking of them as data streams. What we find on the two sides of the Golden Record are two distinct ways of using the data-storage function of the gramophone record. The A-side treats the medium as analog, and the B-side treats it as if it were digital.[21] We humans hear the buzz on the B-side as a kind of music by default, because that is just how our auditory apparatus functions. But there is nothing to say that this is what it necessarily *is*. If we know how to read the signals properly—if we impose a new kind of grammar or band-pass filter on the sound wave—we can get the gramophone groove to record images. It all becomes a matter of finding the right interface.

Kittler, then, got somewhat carried away when he waxed lyrical (or waxed cylindrical) about the gramophone's filterless access.[22] The filter still exists. It just operates on different principles. In fact, it is because the cookie cutter of the gramophone is so well attuned to human hearing that Kittler failed to notice it. But in this interstellar, transspecies context, the filter function of recording technology comes critically into focus—quite literally as an image.

So in summary, the flat ontological surface of the Golden Record is possible because of the data stream of frequencies flowing through its grooves awaiting perception. It is precisely this openness that enables the information on the disc to be accessed in different modalities. What seemingly has no filter is merely an indifference for difference to access. There is always a filter. Interfaces cannot be effaced. To recall an axiom of IMTE, music is universal, but its communication is always particular.

PREMISE FIVE

Music perception exists at the interface, a gap of contingent possibilities that modulate data in the passage from (re)production to reception. The interface functions as a point of disclosure.

0101. A DIFFERENT MUSIC

Exactly why do humans hear the digital data on the B-side as a buzz? In answer to this question, media archaeology and music theory converge on a single page: the flat, continuous glissando of the frequency spectrum from which pitch, timbre, rhythm, and form are folded. To excavate the technology behind the buzz, an archaeology of media would need to dig up some obscure digital noise machines to analyze this spectrum. There are a very few examples of this method of creating music (if we want to call it that on Earth). The acoustical device called the Savart wheel demonstrates the underlying principle in a straightforward way. Anyone who has ever held a coaster or a playing card between the spokes of a revolving bicycle wheel knows how this works. The Savart wheel is a cogwheel that strikes a flexible cardboard in a periodic pulsation and sets it in vibration. Figure 4.7 shows a quadruple Savart wheel that will produce four sounding pulsations simultaneously. When the wheel spins slowly, humans hear a regular pulsation, as a rhythm, and when the wheel speeds up, at around twenty strikes per second or faster, the sensation turns into a pitch. The boundary between pulsation and pitch is called the auditory threshold, around 20 Hz.

Figure 4.7. In the nineteenth century, even popular magazines felt the need to keep their readers up to date with recent inventions in the world of acoustics. An illustration of a four-ply Savart wheel from *Harper's New Monthly Magazine* (1872).

The idea behind the Savart wheel is so simple and compelling that it was discovered again and again at various points in human history: Félix Savart, a nineteenth-century physicist, was hardly the first. The natural philosopher Robert Hooke, who was excited to have identified the effect in the seventeenth century, explained to his friend Samuel Pepys that he could count the flapping of the wings of a fly simply by listening to the pitch of its buzz.[23] It is easy to see why human curiosity would be tickled over and over again by this phenomenon: the means of sound production is visibly discontinuous, whereas the sounding effect is a continuous sensation. At a time when the correlation between frequency and pitch was just a theoretical fact, but had not been shown empirically, the importance of counting vibrations was not to be underestimated. (Just thinking of the fly certainly gave Hooke a digital buzz.) If the rotation speed of the wheel and the number of cogs were known, frequency could be plotted against the sounding pitch.

Electronic and tape-based music has long known of this phenomenon.[24] The composer Karlheinz Stockhausen explained the underlying phenomenon in vivid terms:

> I recorded individual pulses from an impulse generator, and spliced them together in a particular rhythm. Then I made a tape loop of this rhythm, let's say it is tac-tac, tac, a very simple rhythm—and then I speed it up, tarac-tac, tarac-tac, tarac-tac, tarac-tac, tarac-tac, tarac-tac, and so on. After a while the rhythm becomes continuous, and when I speed it up still more, you begin to hear a low tone rising in pitch. That means this little period tarac-tac, tarac-tac, which lasted about a second, is now lasting less than one-sixteenth of a second, because at a frequency of around sixteen cycles per second is the lower limit of the perception of pitch, and a sound vibrating at sixteen cycles per second corresponds to a very low fundamental pitch on the organ.[25]

We needn't worry too much about the fact that Stockhausen sets the auditory threshold at 16 Hz, as opposed to the more common 20 Hz; these figures have always been approximations. What matters is that he included this effect in his electronic composition *Kontakte*

(1954). Figure 4.8 shows a short excerpt from the graphic score for the piece. Sad to say, NASA neglected to include this important piece on the Golden Record, but we can always justify consideration of this piece in this interstellar context with reference to Stockhausen's claim that he was born on Sirius. The way the principle is employed in *Kontakte* is exactly the other way around, as a gradual process of deceleration: we hear a high-pitched sound that meanders and slows down, descending in pitch, until finally we hear it as a series of clicks.

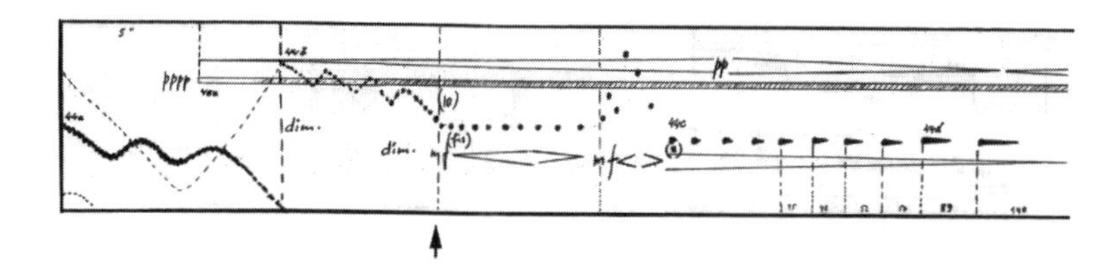

Figure 4.8. A short excerpt from Stockhausen's graphic score for *Kontakte*. The line that starts at the top left as a thin dashed line and turns into a meandering dotted line in the center of the excerpt represents a stimulus that is first played so fast that it is heard as a pitch (dashed line) and then gradually slows down to become recognizable as clicks (dots). The upward-pointing arrow at the bottom marks the moment when the auditory threshold is passed.

Stockhausen's description brings into focus an important principle of data storage for which information technologists have developed a certain fondness: recursion. Simply understood, the principle of recursion is the mise-en-abyme, the mirror in the mirror, cast either as an infinite loop or as a recurrence of the same at a higher level. The dimension of pitch, in this model, is recursively related to the dimension of rhythm. The way the auditory threshold is highlighted in this example is remarkable, because we rarely get to experience it. We can hear Stockhausen's recorded sound object "tarac-tac" as a rhythm or as a pitch — the difference is merely one of timescale. The distinction between the two perceptual parameters, rhythm and pitch, is absolutely fundamental for human music and is dictated by the constraints of human physiology and the way our auditory system operates. Stockhausen is not the only musician to work with this phenomenon. A very different example that works in the opposite direction of Stockhausen's glissando is Moby's electronic dance music track "Thousand" (1991), which was entered into the Guinness Book of World Records as the fastest dance track in history. It toys with the auditory threshold by gradually accelerating the beat until it reaches 1000 beats per minute (=16.6 Hz), which can almost be perceived as a very low, buzzy pitch. The two parameters occupy the two main axes of our notational system, but in the sound wave, they exist in the same dimension, just in different registers of the *t* axis. Another way of saying this is that our human ears fold the sound wave in such a way that different regions of the temporal axis become separate parameters.

The term "fold" here is being used to express the transformation that the sound wave undergoes as it is pressed into multiple dimensions by the auditory system, as one would fold a piece of flat cardboard into a three-dimensional box. It is a kind of recursive origami. There is a strong resonance with the "flattened" Earth Music, stripped of its cultural context and pressed, irrespective of its provenance, into the flat ontology that prevails in the grooves of the Golden Record. Thus stored, it is reduced to its bare essentials — frequencies — before it can be built up again in multiple dimensions, following the "folding" plan of whatever being's perceptual apparatus is found at the receiving end.

This recursive approach to music was doggedly pursued by Friedrich Wilhelm Opelt, a nineteenth-century polymath from Saxony now best remembered by a lunar crater named after him and his son. In our media-archaeological dig into the mechanics of (non) human hearing, Opelt would hold reasonable claims to the title of being the first exomusicologist — and not only because of his lunar connection. Opelt worked as a tax collector by day and dabbled in his spare time in applied mathematics across a dizzying range of fields. Not only did he calculate the diameters of lunar craters, develop an unrealized state-wide pension scheme, and translate an influential mechanics textbook from the French, but he also developed a recursive music theory based on precisely the mechanisms we are discussing here.[26]

Opelt grounded his speculations in a noisemaker that was the equivalent of the Savart wheel, but one that had the additional advantage of working just with air: the mechanical siren. Presented and named by the French engineer Charles Cagniard de la Tour in 1819, the original siren, shown in Figure 4.9, consisted of a metal disc with a series of holes punched out at regular intervals. The bores of these holes are diagonal, and they sit on another disc with diagonal bores in the opposite direction. When air or another fluid is blown through the bottom disc, the top disc is set in rotation. (Cagniard de la Tour chose the name *sirène*, or "mermaid," because his device can "sing" in air and water.)[27] Every time the bores in the two discs coincide, an air puff is emitted, producing a series of on/off impulses. The higher the air pressure, the higher the rotation speed and the louder the noise emitted by the siren. At low rotation speed, human ears hear a regular rhythmic pulsation, and when it speeds up, at the auditory threshold, the pulsation turns into a pitch—exactly as in our previous examples. The rising and falling glissando produced by changing rotation speed gives the siren its characteristic sound.

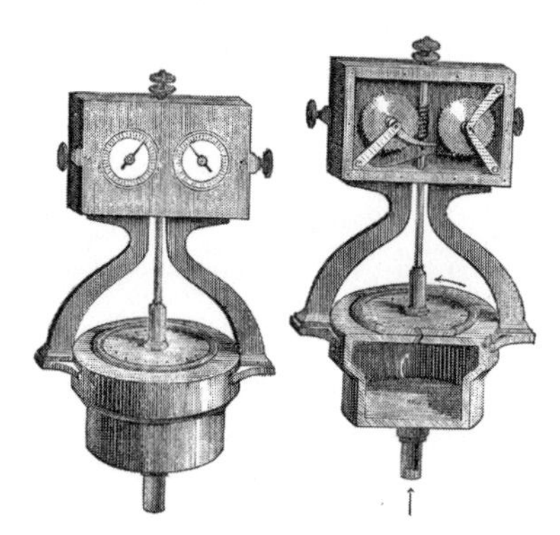

Figure 4.9. Cagniard's siren, from *Harper's New Monthly Magazine* (1872). The right image shows a cross section of the mechanism, with arrows indicating the air flow and rotation of the metal disc. The counters at the top of the apparatus indicate the number of rotations, which makes it possible to calculate the frequency.

The siren is an apparatus that can be used to analyze (human) hearing. This thought takes us to the heart of our reverse engineering exercise and to the bottom of our media-archaeological dig. It lays bare the machinations of the ear by building an apparatus that contrasts the sound-production mechanism with the aural effects it produces: what our eyes see does not match what our ears hear. By highlighting those differences, we become more aware of mechanisms that are at work all the time, but usually pass unnoticed. Put differently, this is why an archaeological dig into the media landscape of past times can paradoxically help us think about the future.

Opelt found that it is possible to produce multiple simultaneous sounds in two ways.[28] First, as was widely known, many more rings can be punched into the siren, with each one creating its own sound. And second, as Opelt found, different pulsations can also be superimposed on the same ring. The compound rhythm of these multiple pulsations, even if punched into one single stream, will produce multiple sounds, with the intervals corresponding to the ratios of the pulsations.

Opelt's siren operates as a sonic weaver, spinning threads of frequencies from its algorithm of tiny holes and twisting them in complex binary patterns. To take a simple example, a hemiolic rhythm—triplet against duplet—can be combined into the compound rhythm *dum-di-di-dum dum-di-di-dum* on the disc. As Pythagoras and his hammers taught all those years ago, this ratio (2:3) corresponds to the interval of the fifth. Opelt's siren demonstrates audibly that this was never just abstract cosmic hocus-pocus: given sufficient rotation speed, the siren sounds a fifth, which rises in pitch as the disc rotates faster. More complicated compound rhythms work in similar ways. Rudolph Koenig of Paris, a prestigious purveyor of nineteenth-century acoustical instruments, produced a sample siren disc based on Opelt's ideas, shown in Figure 4.10, that goes through various combinations, producing an intricate symmetrical pattern of holes, of presences and absences.

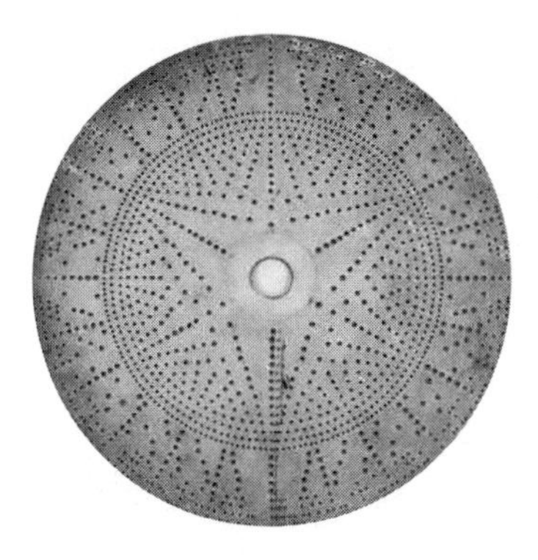

Figure 4.10. Opelt's siren. Each circle represents one combination of patterns that can be perceived as either pitches (intervals and chords) or rhythms, depending on the rotation speed. *Left*: an Opelt siren disc made of cardboard, produced by Rudolf Koenig (Harvard University, Historical Collection of Scientific Instruments). *Right*: a demonstration of the Opelt siren, powered by air blown out of a tube, from *Harper's New Monthly Magazine* (1872).

The lacy patterns that make up Opelt's siren constitute a musical writing by hole punch. It is a media reversal of the Golden Record's B-side: a digital image of sound. Opelt's circular notation spatializes temporal relations and stores sound in visible form. The more complex and irregular the circular punch-hole pattern, the more voices we hear when the rotation speed crosses the auditory threshold. (The outermost loop of the disc in Fig. 4.10 can sound a four-voiced chord.) His siren presents an object lesson of the mise-en-abyme principle of recursion: the hemiolic *dum-di-di-dum* rhythm *is* the interval of the fifth. The two are identical. The only difference between the two renditions is the playback speed—the same motive performed *molto lento* or *vivace assai*.

Opelt used his looping system to drive this idea of the recursive relationship between musical parameters as far as he possibly could, including not only pitch and rhythm, but also harmony and phrase rhythm. In one sense, it was obsessive; it is an overdetermined system that literally spirals in on itself.[29] But by pushing its logic to the limit, his theory reveals some fundamental components of music. Opelt's significance for our exomusicological expedition is twofold.

1. Opelt's machine demonstrates how the flat surface of the frequency spectrum folds. The parameters generated by the machine—timbre, pitch, rhythm, and form—are all related across the same temporal axis.[30] As formalized in the IMTE blueprint, we can ventriloquize for Opelt that rhythm occurs predominantly in the simple Hertz (10^0 Hz) range and pitch in the kilohertz (10^3 Hz) range, whereas form happens around the millihertz (10^{-3} Hz) range—and, we might add, timbre, if we go further into the higher kilohertz range. The constraints of the human ear folds an indifferent fabric of frequencies into these specific pleats. In other words, the frequency spectrum is not music per se, but the site for music. Where there is recursion on the frequency spectrum, there is the possibility of music, because recursions enable repetition to occur as complex, internally generated patterns.
2. Opelt's rotating machine, with its alternation of blocks and holes, digitizes the frequency spectrum. It maps the binary basis of music—rhythm—and how its complex repetitions and recursions weave a frequency of overlapping and intersecting patterns.

In this context, it is more than a poetic accident of history that Opelt came from a family of weavers. He is a "Penelopean" theorist of the industrial age. The mechanical Jacquard loom, which revolutionized weaving manufacture in the early nineteenth century, produced intricately patterned fabrics en masse. The mechanism that made all this possible was based on a system of punch cards in which the patterns were stored and translated into mechanical action. Using a lever system, the warp was raised and lowered and allowed the shuttle to whiz across the loom with unprecedented precision and complexity. The Jacquard loom has been seen as a forerunner of the computer.[31] From this angle, Opelt's siren discs are nothing other than circular punch cards computing sound. Both the Jacquard loom and the siren are mechanical versions of media that store and transduce a digital code into different sensual modalities.[32] The warp and weft of Opelt's musical loom are the specific constraints of the human auditory system.

Which is why, for all the materialist posturing, IMTE is not fully explained by the medium that carries music from our blue planet to another star system. It is a string theory of dangling strands. The sender has delivered a musical fabric from its own loom, but what is being transmitted are merely the strings, the vibrations, the flying shuttle of Earth sounds — it does not contain the loom. The warp needs to be thrown anew at the other end, and the weft must be woven around it again by the recipient. Who knows what the tapestry at the other end might look — or sound — like.

PREMISE SIX

Exomusicology recognizes the recursive folds of human perception. To understand the alien interface, it removes the boundaries between parameters shaped by human perception and exposes the underlying glissandos in the data to rebuild the possibilities of reception. To determine where music might be relocated on the frequency spectrum by an alien audience, the recursive properties of the music must be mapped in relation to the recursive configuration of the alien's receptors.

0110. THE BUZZ

Returning to Voyager and the buzz whirring at 30 Hz on the B-side of the Golden Record, to human ears, these images are located above the human auditory threshold as a hum somewhere between low B♭ and B. But they are also clicks, like Stockhausen's "tarac-tac," that may be perceived as different things if different kinds of ears hear them. So to abstract a little more, we should home in on the question of whether Opelt's sound-generating mechanism is analog or digital.

The answer depends on which way we look at it. The siren generates sound by means of air puffs punctuated by blockage, that is, an alternation between air and no air. This description, which focuses on presence and absence, clearly falls into the digital realm. But at the same time, the boundaries of each impulse are reducible — as all wave forms are — into sinusoidal components, according to the rules of Fourier analysis. Analog, after all, is not the opposite of digital, but its twin. This dual nature of the buzz translates into the cross-modal demands that the B-side makes on its audiences. The sonified images are a Necker cube of human perception.[33] We know the sonified images are intended by the makers of the Golden Record as pixelated and digitized visual information that translates into dots within a matrix, but operating within the human sensorium, we cannot help perceiving them as sound, because everything recursive is music to our ears. Technically, what we hear is not the image itself, but the rasterizing process that makes these images temporal as they are being scanned from side to side. Our ears are hearing the temporal filter for the digital image as analog sound.

Media theorists have long pointed out that technical media store temporal processes by spatializing them,[34] but the sonified images, in turn, temporalize a spatial phenomenon. The image takes time to appear and cannot be grasped as a fixed instant. By bending the "time's arrow" of the sound wave twenty-nine times into equal durations, the individual lines of each pixelated image emerge across the matrix, and in this way, the sound wave can fill the contours of each image with dots and spaces. As humans, we have to reconstruct these images mechanically, most expediently by using a computer to assist in this task. But the computer is an interface that transduces and sculpts frequency data into an expressive form, just as our sensory apparatus does at every waking moment. The folding of short beeps into a regular 29 × 19 matrix is not fundamentally different from the recursive folding that our ears habitually perform on the sound wave. Would it be beyond the imagination that a different auditory system might exist in a distant corner of the galaxy that corresponds to the twenty-nine folds of the sound wave that recreate each image on the record?

This would be an exceptional stroke of good luck. But even if we banish such an outlandish sensory modality into the wishful realm of science fiction, the B-side of the Golden Record can teach us to align our conception of music with data analysis by turning music into digital code. What remains are mere vibrations. These vibrations must be reconfigured to fit the auditory sensory apparatus of whatever extraterrestrial being will pick up the Golden Record at the other end. This is the interface that will carry us beyond where the stylus left us dangling, right at the abyss into the unknown.

To transmit Earth Music to another planet, there is one final step that we need to take in our media-archaeological journey into the future. If recursion describes the gesture by which the selfsame is repeated at different levels, then Voyager represents music's ultimate recursion. As Wolfgang Ernst asserts, a lyre that was struck in ancient Greece two thousand years ago would sound *exactly the same* as a lyre struck now.[35] This may seem a bewildering claim, because there are clearly significant cultural differences between Greek antiquity and our modern world. Ernst's proposition is true only at a molecular level; the pressure waves in the air travel from the instrument to the perceiving ear following the same natural laws at all times. As soon as we get beyond this elementary stage, however, the differences become so vast that any similarity, let alone identity, rapidly evaporates. But this is also precisely the point of such historiographical recursions: the physical commonality between both events, divided by chronology, provides the material basis of comparison; it is the divergences between the two, as we get into the messiness of reception, that become telling.

The selfsame sounds, in a state preceding signification as mere frequencies, become a wormhole that connects vastly different periods and places, enabling different modes of reception at either end. In the case of the Golden Record, of course, we are not talking about a piddling time span of two millennia and the negligible geographical distance between Athens and Berlin. But the principle holds, no matter whether the vessel in which the sounds are stored is classified as a lyre or a spaceship.

What has traveled the distance is the raw data; this allows the comparison. We cannot know how it will be folded, how it will be interpreted, and what filters will be applied at the other end, but the presence of a constant makes divergence possible. Frequency is this constant; perception is this divergence.

PREMISE SEVEN

In exomusicology, recursion also describes the recurrence of the same frequencies across vast distances in time and space. The Golden Record is a wormhole that delivers the data. The frequencies are the same; they converge at both ends, but the reception will be divergent.

The sonified images on the B-side of the Golden Record constitute an object lesson for exomusicology because they confront us with a different kind of cookie cutter to which human ears are insensitive. Or, in keeping with the weaving metaphor, these images thread a different warp and weft on the loom of the auditory apparatus. The same is probably true for the samples of music on the A-side when presented to nonhuman ears. We cannot know, let alone hear, the specific constraints of nonhuman perceptual systems. The options are wide open. If we want to stare into the great unknown that lies beyond the stylus, we have to learn to let go. We have to let go of ears.

But this letting go is not a dispirited emptying of knowledge. Rather, we move forward by becoming more attuned to the way in which interfaces operate on music, whether these are computing machines or sensory organs. We get furthest in this interstellar context if we go back to basics and return to music as data streams and vibrations. These vibrations must be reconfigured to fit the sensory apparatus of whatever extraterrestrial being might pick up the Golden Record at the other end. This is the interface that will carry us beyond the point where the stylus left us dangling, right on the edge of the abyss that leads into the unknown.

To summarize: to theorize the unknown, we must theorize our constraints. To understand nonhuman modes of listening, then, we must begin with an awareness that the multiple dimensions of our sound world are all temporal functions riding along the sound wave, each with their own limitation. Even though we perceive them as discrete parameters, our multidimensional perception is a radical interpretation that our sensory apparatus (an interface) imposes on the sound wave. Our ears fold the sound wave into origami, just as the Golden Record instructs the unknown alien to bend the image recordings into twenty-nine lines of equal length and weave them into a two-dimensional matrix. We cannot determine the unknown, but one thing is for sure: if the stylus leaves us at the abyss, the folding of the sound wave in accordance with other filters and interfaces will take us across.

THE SEVEN PREMISES OF ALIEN COMMUNICATION

To close, we present the seven premises dotted sporadically across the chapter as a summarizing sequence that outlines a process for alien communication, from numbers, to frequency, to vibrations, to recursions.

1. Numbers are the medium of communication with other intelligent life-forms across the universe.
2. Numbers are measured as frequency—a binary computation of similarity and difference. Frequencies are a form of counting and comparison that can be stored in a machine as data.
3. Vibrations are the concrete expression of numbers and are played back as an event at the point of contact.
4. Given the vast timescale envisaged in exomusicology, media archaeology is a necessary method deployed to imagine the survival of music after humanity in terms of its storage as frequency and its playback as vibration. Media archaeology reverse engineers the future to determine the conditions necessary for alien communication.

5. To communicate to an alien other, exomusicology theorizes the interface where musical perception occurs. This point of disclosure is a gap of contingent possibilities that particularizes reception.
6. Exomusicology recognizes the recursive folds of human perception in order to remove these boundaries, exposing the underlying glissandos in the data to rebuild the possibilities of alien reception. To determine where music might be relocated on the frequency spectrum, the recursive properties of the music must be mapped in relation to the recursive configuration of the alien's receptors.
7. Finally, recursion also describes the recurrence of the same frequencies across vast distances in time and space. The Golden Record is a wormhole that delivers the data. The frequencies are the same; they converge at both ends, but the reception will be divergent.

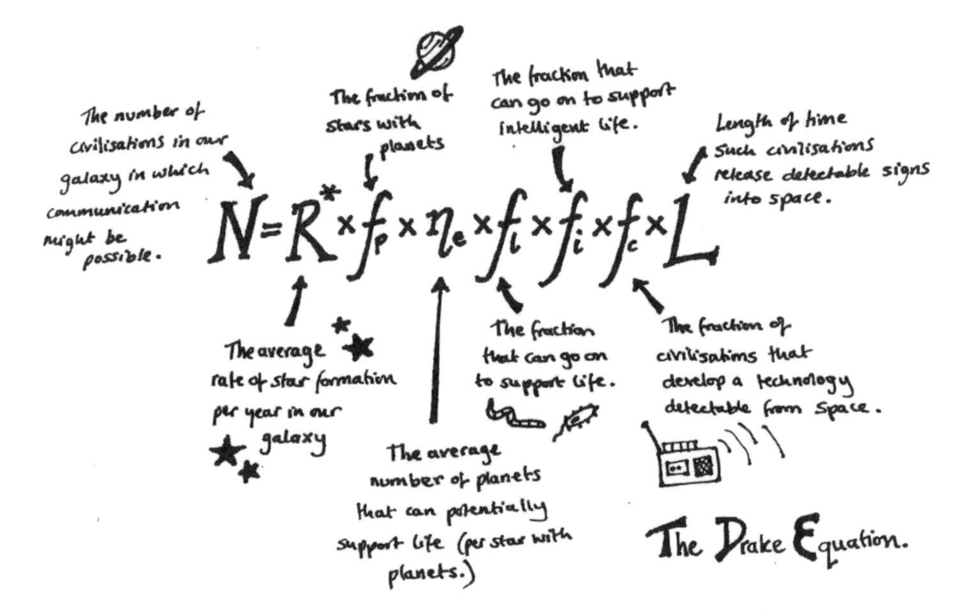

Figure 5.1. The Drake equation as internet meme.

CHAPTER FIVE

Receiver

There be whales here.

—*Star Trek IV*

$$N = R^* \cdot f_p \cdot n_e \cdot f_l \cdot f_i \cdot f_c \cdot L$$

This is the formula that is used for calculating the number of alien civilizations in the Milky Way. It was devised by Frank Drake in 1961 to stimulate discussion at the first meeting of SETI. Figure 5.1 explains the different components of the equation. Sixteen years later, Drake would test his own equation as part of the curatorial team that rocketed the Golden Record into space. Despite appearances, there is nothing mathematically precise about the Drake equation, because each of its variables is conjectural. With each variable, the formula zeroes in on the specifics of communication, becoming progressively improbable as it homes in toward the "Goldilocks planet." Such a planet cannot be too large or too small, it cannot be too far from or too close to its star, it cannot have too much or too little water, and so on. In order to sustain life, it must be "just right."[1]

It is all very demanding. In fact, the Golden Record is even more exacting than Goldilocks, adding a further dimension of complexity to the equation. It presupposes human attributes for aliens, including extremities that can handle the record and construct a gramophone and an auditory system with which to process the vibrations. This is because Carl Sagan optimistically assumed that alien civilizations must be technologically superior to us—that they can do everything humans can do, and then some . . . then even more. Sagan's aliens should have no problem with the Golden Record. They are "human+" beings. Even if these extraterrestrials were, say, so hot that the Golden Record would melt on contact or so small that they could only wriggle inside the grooves, Sagan's firm belief in science and technology ensures that they would overcome any obstacle between them and the golden disc.[2]

So given the need for such hyperintelligent beings, alien contact with the Golden Record is highly unlikely. The odds are longer than Pythagoras's cosmic string. Among the billions and billions of exoplanets, only a few will qualify as recipients of the Golden Record. But, at least, the chance is nonzero.

In interstellar communication, it is easier to give than to receive.

0001. SPECIES COUNTERPOINT

So far, the search for extraterrestrial intelligence undertaken by SETI has come up with nothing. There is only a gap. Despite the increasing belief in the possibility of alien life beyond our star system and the myriad of radio telescope dishes twitching like giants ears to hear an incoming call from outer space, there has been no contact. We are still alone. Of course, the awkward silence between the probability of extraterrestrial life and the lack of contact—known as the Fermi paradox—will take only one instance to be broken, but even if an alien were to dial in, what on earth would we beam back? It is easy to cut SETI down to size.[3] The ambition to communicate with extraterrestrials is quickly deflated by pointing to the awkward fact that humans have hardly managed to communicate intelligently with life on our own planet. Dolphins, whales, and octopodes are, by all accounts, highly intelligent species with their own modes of communication. Why not start here on Earth, the criticism goes, before we move on to more ambitious intergalactic goals?

SETI may not have successfully cracked the codes of nonhuman communication, but exomusicology can accept that challenge for the comparatively simpler task of imagining alternative forms of audition. And the criticism contains an important kernel of truth: by considering nonhuman and nonstandard forms of audition on Earth, we might gain a deeper understanding of aspects of the human sensorium that we would otherwise take for granted. These defamiliarizing forms, which question our norms for analyzing music, are good test cases for the more ambitious task of communicating with the unknown alien that the Golden Record presupposes. If music, as frequency, is universal, but particular in its communication, then an attempt to hear with different "ears" would make audible the filters and interfaces that transform, rather than transmit communication. By examining auditory differences, what we hear as transparent will become constructively opaque.

Let us begin with humans. Forms of noncochlear human listening are not uncommon. The deaf percussionist Evelyn Glennie has famously written about her form of experiencing music. When she performs, she takes off her socks and shoes, since the vibrations that we encounter as sounds are transmitted via the floorboards and enter her body from the soles up. Other deaf people in history developed similar somatic listening strategies, including Edison and Beethoven, who would bite into vibrating sound objects — the gramophone sound horn and a stick connected with the piano — to make the most of bone conduction. These noncochlear listening strategies are highly dependent on the acoustic properties of the vibrating surface and vastly change the nature of what is being perceived and processed. Lower frequencies, for example, are often exaggerated and tend to boom and explode; distinct frequencies become fleeting, muffled vibrations bereft of place and location; virtual sounds can be triggered in the mind by the movement of objects caught by the eye. The question of whether this listening still constitutes music is easily swept away with a limp wave of the hand in the general direction of musicians with such impeccable credentials as Glennie or Beethoven. Music can still be music, even if the interface radically changes. It is a matter of adjustment, of fine-tuning the reception, of analyzing the difference and restrategizing the process of listening.

Take cochlear implants. This device is a perfect illustration of the filter function that all auditory systems impose on sounding data: cochlear implants can transmit only certain narrowly defined frequency ranges to the brain; any sound that does not coincide with these frequencies cannot stimulate a neural firing and is quite literally filtered out. Where a healthy cochlea has about thirty thousand hair cells that are each tuned to a specific frequency, cochlear implants stimulate only about a dozen positions along the basilar membrane. The devices are particularly designed to replicate speech, however imperfectly, and do not transmit pitch or timbre well. Human music does not resonate well in this system. Some composers have been working on writing music specifically for cochlear implants that capitalizes on the transmitted frequencies, which demonstrates the strategic difference that interfaces make in communication.[4]

Such human-made technology underlines the general fact that the auditory apparatus itself is an interface in our telephone model of sender-transmitter-receiver. But the act of transmission is not only like a telephone; it is also a game of telephone (or "Chinese whispers"): data is never just conducted neutrally, but is always transformed in the process of mediation. It would disarm Sagan's optimistic, but all too human version of alien intelligence—the "human+" being. Following Luhmann's *ego/alter* model, this chapter raises the curtain on a menagerie of nonhuman ears in different sizes, physiologies, metabolic rates, and habitats to highlight the infinite diversity of possible auditions, effacing the *ego* to allow the alienated *alter* to take center stage.

Of course, this menagerie cannot be a comprehensive list of alternative ears on Earth. It is a collection of ears, like the compilation of music on the Golden Record. What it does is zoom in on the f_i variable in the Drake equation, the fraction of planets with life that actually go on to develop intelligent life, from the specific angle of audition. Our representative sample of ears on Earth is an exercise in species counterpoint.[5] It is not only a species counterpoint of different life-forms interconnected within an ecosystem, but also a species counterpoint of possible types of audition from which we might imagine modes of communication. Hearing the other — whether between races, genders, species, or planetary civilizations — is a defamiliarizing act that is the initial step for all alien contact.

0010. MENAGERIE

Chiroptera What is it like to hear like a bat? Pretty weird. The function of bat audition is much closer to what we normally, in our anthropocentric worldview, associate with vision. The ultrasonic shrieks that bats emit for the purpose of echolocation, around 100,000 Hz, far beyond human hearing, which ends around 20,000 Hz, are reflected off obstacles, which allow the flying mammals to navigate their environment by measuring the time delay of the echo. Latency translates into physical space in bat ears.[6] The extremely short wavelengths of these bat signals allow them to scan the surface of objects: in experiments, bats have been shown to eat only real insects and to ignore identical-looking bait made of plastic.[7] The sonic information reflected off the surface of the objects communicates crucial textural differences between true exoskeleton and mere imitation. Bat hearing outstrips even human vision and crosses into the perceptual range that we associate with touch. Their hearing would literally "get into the groove" of the Golden Record. Bat listening would be a tactile "thing." It would not need the stylus provided by NASA: technically, a bat would just need to rotate its head at the right speed to audit the grooves. The bat is its own gramophone.

Bats coevolved with moths. A moth's auditory receptors can sense frequencies up to 300 kHz and are specifically targeted to respond to the ultrasonic shrieks of its predator. What is an echolocation tone for a bat, then, is a warning signal for a moth; moths can hear only danger. The shared frequency between the two species shows how hearing is an ecology, an auditory network of interconnected perceptions and signs. Or as Jakob von Uexküll would put it, the moth and the bat are "connected contrapuntally," brought into relation by the "score" of their biosphere.[8] Human music, in this finely tuned lifeworld, would be out of sync with their coevolved counterpoint. So although moths have among the sharpest hearing of all the animals on our planet, playing the Golden Record would be an entirely silent performance for the moth . . . unless it was played by a bat, in which case, it had better hide.

These are just playful projections. Ultimately, it is difficult to describe the sensory world of species that have vastly different experiences from us. Thomas Nagel concluded his iconic 1972 study of bat consciousness, "What Is It Like to Be a Bat?," with a shrug of the shoulders. We cannot really know. This is because as with SETI's attempt at a long-distance relationship with aliens, we have not yet learned to communicate with bats.

Nevertheless, by applying strategies used in nonhuman studies—actor networks, flat ontologies, assemblages, animal lifeworlds, and other mixed systems of objects and species—a field is laid out in all its potentialities that circumscribes the human by studying its various others.[9] In the process, the human loses its accustomed state of exceptionalism and gains a new perspective through the recontextualization. Its *ego* is decentered by the *alter*. Such an exercise in alterity will always be inadequate, but necessary. Even the awkward reference to metaphors and similes—"nonhuman listening is a bit like X"—which may appear self-referentially makeshift, is a necessary part of exploring the world beyond our sensorium. We have nothing but the constructs of language and our past experiences to measure difference.[10]

Cephalopod The boldest construct of such an *alter* experience must be Vilém Flusser's study of the vampire squid, *Vampyroteuthis infernalis*, a deep-sea cephalopod.[11] It resides in utter darkness, in the depths of the sea as

if it were floating passively in an expanse of Cagean music; its entire body, with its eight tentacles, three penises, and seventy-five thousand teeth feels the rhythm of sea currents and the drift of "marine snow" (plankton, feces, disintegrated carcasses), occasionally creating its own visual universe with a burst of bioluminescence to dazzle its prey or sexual mates. In Flusser's "fable," which, like any fable, takes an animal to hold a mirror up to human behavior, the vampire squid assumes the role of a nonhuman other, simultaneously disgusting and fascinating, equally developed as *Homo sapiens*, but operating in a physiology, collective consciousness, culture, and ethics diametrically opposed to what we generally espouse as human(e). The values that Flusser's cephalopodology extracts from the handful of facts that we know about this rare animal can get quite dark at times—but only when judged from an anthropocentric perspective, which is precisely what he is at work decentering. Whether these speculations and extrapolations correspond to any cephalopodological reality is unclear; they matter only insofar as they represent a value system that steps outside the categories that our anthropocentric attitude imposes on our world.

There is in fact a good case to be made that cephalopods would make for an excellent model of nonhuman consciousness—the closest we would come to an alien on Earth.[12] What is remarkable about octopodes, a highly intelligent species, is the decentralized organization of their nervous system, which is unlike anything else among earthly life. Each arm has a certain degree of autonomy in ways that we, constrained as we are by the straightjacket of our central nervous system, can barely begin to imagine. Not much is known about octopus audition, but we know that they can perceive sounds between roughly 400 and 1,000 Hz, a very limited frequency range. Translated into human terms, this is from about G4 to B5; the octopus's "obligatory register" covers barely a minor tenth.[13] The entire upper range of the piano keyboard would be inaudible to octopodes, much as dog whistles are to human ears.

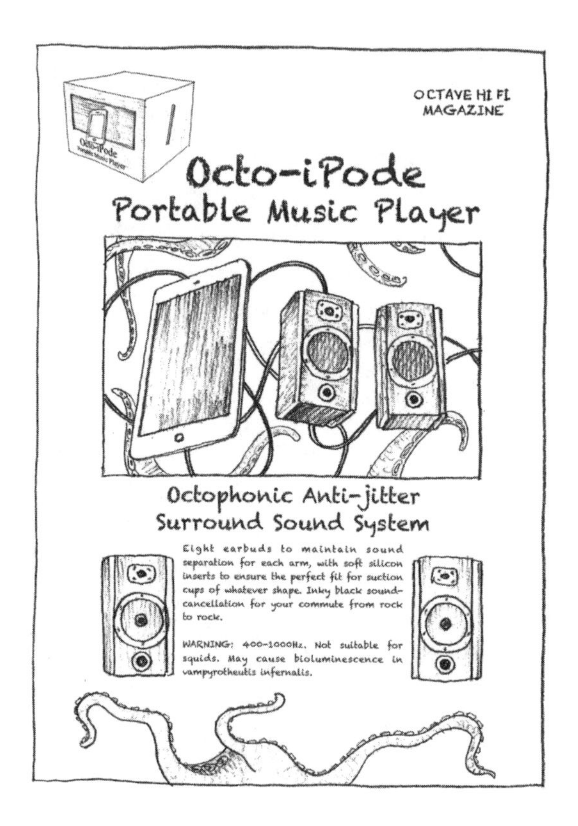

Octopodes do not have ears. They use their statocysts, organs that help them stabilize and establish a sense of balance, equivalent to the semicircular canals of the human inner ear, for a form of hearing. Octopodes' statocysts are lined with hair cells. If octopus hearing follows the same basic principles as human hearing—and for now, we have to leave a big question mark hanging over this question—then this high auditory threshold would mean, given the recursive relationship between rhythm and pitch we identified using the Savart wheel, that octopodes have exquisite rhythmic perception, identifying minuscule time intervals that have long fused into a pitch for human ears. An octopus might hear frequencies below 400 Hz as rhythmic pulsations.[14] That's 24,000 beats per minute. Moby's dance track "Thousand" would have to ramp up its beat track twenty-four times as fast before it could achieve a remotely similar effect in octopodes. It also means that the opening theme of the Fugue in C from J. S. Bach's *Well-Tempered Clavier*, book 2, on the Golden Record, which ranges from C4 to A4, would largely be a rhythmic gesture in the octopus world.

But we should not take pitch for granted. We have already seen that the capacity for pitch is an effect of the thirty thousand or so hair cells of the human ear. There is evidence that even our very close evolutionary relatives, rhesus macaques, do not share certain features of our auditory system.[15] It seems that apes are less sensitive to specific pitches and more attuned to noise. This might mean that the central feature of human music perception—the ability to fuse a complex timbre down to a single pitch, which is usually heard at its fundamental frequency—plays little or no part in ape audition. What human ears hear as the solo melody of the Peruvian wedding song on the Golden Record may sound like a complex polyphonic texture of individual upper partials to macaque ears. It is not inconceivable that for nonhuman primates, all music sounds similar to how throat singing sounds to humans, in which certain prominent overtones ring out above the general complex noise, with no sense of a unified blend coinciding with the lowest harmonic.

To return to our octopus model, it is not just pitch that needs estranging; it is equally questionable to what extent "hearing" is the right word for the sensory perception of the cephalopod: in all likelihood, the hair cells in their statocysts that detect fine-tuned motion are closer to a jiggling Jell-O than to what we generally associate with hearing . . . which makes our assumptions about octopus audition somewhat wobbly.[16] What the octopus receives in the frequencies we call pitch would be utterly alien to our ears. Indeed, we would not even share the same sense of time. Octopodes have a much higher metabolic rate than humans, and there is some evidence of a correlation between metabolic rate and temporal perception.[17] The flicker fusion frequency (FFF)—the visual equivalent of the auditory threshold — is much higher for certain animals than it is for humans. That's why it's a bad idea to take a pigeon to the movies. The speed at which the movie is played, at twenty images per second, is good only for human eyes. The bird's FFF is above this threshold, and the bird would just perceive a slide show.

Drosophila This idea was taken to its logical extreme by the nineteenth-century biologist Karl Ernst von Baer, who conceived the thought experiment that our experiential world is a function of our temporal experience. Each life-form, he claims, is organized by its own internal rhythm.[18] Assume an animal, most likely an insect such as a fruit fly, that lives only for one thousandth of an average human life span of eighty years — that is, about twenty-eight days. This insect, however, has the same number of heartbeats as that average human, which is von Baer's way of saying that its metabolic rate is a thousand times that of a human. The assumption goes, in other words, that the total number of experiences would be the same for both species, but greatly compressed in the case of our short-lived insect. Each sunrise, each sunset, would be a major life event. Since its life span would be shorter than a lunar cycle, the animal would be able to observe only a shape-shifting object moving across the nightly sky, but would be unable to deduce any regularity. But if the insect's rate of perception is so much higher than it is for humans, presumably, what we perceive as very high-pitched sounds would be heard as very low sounds. Edda Moser, flaunting her stratospheric voice as Queen of the Night in Mozart's *Magic Flute* on the Golden Record, would come across as a *basso profundo* to our tiny, ephemeral creature. The track time of two minutes, fifty-five seconds would be the human equivalent of forty-eight hours, which would be a considerable operatic commitment for a fruit fly.

Cetaceans Marine mammals, especially whales and dolphins, have long captured the imagination of interspecies communication. Since the 1960s, scientists have engaged in acts of wishful anthropomorphism, imposing human frameworks of music and language on their clicks and grunts. Reimagining their phonations as songs virtually transformed cetaceans into our intellectual soul mates. In 1970, the record *Songs of the Humpback Whale* made the music of baleen whales widely known among human listeners. Or more precisely, it cemented the idea in human consciousness that baleen whale calls constitute a kind of music. The parallel is not implausible: scientists have subjected their calls to music analysis and categorized them into song types and song forms; they have pointed out the hierarchically organized nature of the sounds, with repeated units structuring complex sequences and themes.[19] Moreover, the parallel with human musicking extends to their method of distribution: male humpback whales seem to "compose" new songs on an annual cycle, some of which are of epic proportions lasting up to thirty minutes; these are then repeated and transmitted across different population groups, like a vast oceanic game of telephone.[20]

For humans to access whale song, the water through which the sound travels has to be "filtered," since most underwater vibrations would simply bypass our eardrums: any direct contact with a singing whale would just rattle our bones and shake the body as a monophonic shudder devoid of spatial cues.[21] Whale songs were widely distributed for human reception via the hydrophone, through which their voluminous frequencies become shrunken objects, heard from a distance and seemingly muffled by vast stretches of "reverberating" ocean. A whale may not recognize its own music through this interface, but this technological intervention connected an alien world sonically to our airy environment, making the 1970 recording of the humpback whales somewhat like the Golden Record: humans had discovered an intelligent alien species through music. An excerpt from this album even makes a brief appearance on the Golden Record during the mashup of the UN delegates (to avoid any whiff of provincialism, as Sagan emphasized).[22] Not coincidentally, the whale recording played a major part in the success of the "Save the Whales" campaign that banned commercial whaling—rebranding their calls as music was crucial to humanizing the endangered cetaceans.[23]

But even before the song endowed the whale with emotional intelligence, cetaceans were celebrated as an emblem of linguistic intelligence. SETI is directly related to this cultural elevation of cetaceans: the group's original name, the Order of the Dolphin, is an oblique reference to the possibility of communicating with these intelligent nonhumans.[24] Cetaceans were regarded as oceanic extraterrestrials, their watery cosmos mirroring outer space. At the time, NASA even experimented with human-dolphin cohabitation, leading to questionably intimate close encounters in the attempt to understand their language.

There is a subtle slippage between these two positions: *Songs of the Humpback Whale* imagines cetaceans as a music-making species, whereas the Order of the Dolphin regarded them as a speaking species, predicated on the potential to translate their shrieks and clicks into human code. But from Sagan's perspective, this slippage was not a problem for the Voyager mission; the ambiguity is fully intentional. Sagan's approach was a riff on the trope of music as the language of emotions. This is speech expressed from the inside out, a music that could bind species in sympathy, as the ban on commercial whaling evidently demonstrated. Thus, the whale song functions as both a language and the soundtrack in the UN linguoscape on the Golden Record. It is both message and feeling. But despite the artistic aspiration, this interspecies recording is more a confusion than a fusion, conflating differences between two species that have yet to communicate to each other. The whale song, wallowing lugubriously in the background, may bestow an aura of aesthetic cohesion to the collage of human messages, but we are not the same.

For a start, the sea has long been recognized as the big other on Earth. The assumed equivalencies between humans and cetaceans belies the radically different lifeworlds that they inhabit. Their entire physiology is attuned to their watery environment in a way that makes little sense to humans. To communicate effectively underwater, a whale's vocal mechanism is designed to blast its messages at decibel levels that can outperform any heavy metal concert and rupture our eardrums. Dolphins and whales, with their lack of feet (and impermanent dwelling places) and of hands (and, hence, a slippery relationship to objects), are veritable aliens on our blue planet.[25] Ironically, given their fins and flippers, one of the most intelligent species on the planet cannot handle the Golden Record. It would literally slip out of reach. Cetaceans cannot access the music as a thing; data storage blocks communication. But if someone played the record for them as an underwater event, their acute sense of hearing would register the music as a massive and immersive surround-sound experience across vast distances. Humpback whales can pick up calls across the ocean from seaboard to seaboard, and odontocetic whales and dolphins orient themselves by means of ultrasonic echolocation, comparable in their acuity with what bats can accomplishments in the air. Their massive brains would process the sounds on the Golden Record as multisensory images of the oceanscape. But most importantly, the liquid medium they inhabit has excellent sound-conducting qualities, which makes bone conduction an advantageous listening strategy. Cetaceans effectively hear with their jaws.[26]

Which makes whales rather like Beethoven or Evelyn Glennie.

OOII. MIND THE GAP

Our aural examination of various species with radically different auditory ranges (bats), physiological mechanisms (octopodes), psychophysical makeup (macaques), temporal perceptions (birds and insects), and habitats (cetaceans) demonstrates how hearing is a multisensory process embedded in a complex, interconnected network of mediations. Hearing is distributed as an ecosystem, or, in Uexküll's terminology, a symphony of contrapuntal intricacies in which point and counterpoint perfectly align.[27] An object such as the Golden Record needs to plug into the system. But, given the infinite possibilities of hearing latent in the universe, the chances of a workable connection are slim. Even our very limited menagerie of examples already covers a dizzying array of auditory apparatus, listening strategies, and ec(h)osystems. These nonnormative kinds of Earthbound hearing present only a tiny subsection of what is out there beyond our blue planet. What the Drake equation demonstrates is that the greater the necessity for the fine-tuning of the right conditions for reception, the greater the possibility of disconnection.

The Greatest Engineering Failures in the Universe
#1: The Cross-Galaxy Tunnel

Exomusicology must begin from this glitch. Wherever there is species counterpoint without coevolution, transmission and reception are two tunnels that do not necessarily connect. Their misalignment leaves communication hanging precariously. Since music in space consists of vibrations that are communicated via interfaces that always particularize the transmission differently, we are left to theorize the gap. Miscommunication must be an underlying premise of exomusicology as long as the auditory system of our alien recipient remains the big unknown. So in our attempt at species counterpoint, there is more "counter" than "point."

But, more importantly, the gap is not empty. It is not void. It is bristling with infinite possibilities. It is more potential than negation. Far from shutting down the conversation, the premise of miscommunication generates a creative risk—an imaginative leap into the abyss. Voyager's Golden Record is such a risk. In one sense, it is an utterly unnecessary mission, a lavish waste of public funding that in all probability will end in failure. On Earth, species take such flamboyant risks only when sexual selection overtakes the urge for natural selection. Voyager's piece of space bling, flashing in the starlight, is a display to seduce the unknown other. From its golden surface to its musical frequencies, the record craves attention precisely because there is a gap; it overcompensates to overcome the improbability of communication. It blinks, it winks, it repeats itself to draw attention to its existence. In this sense, NASA's Voyager mission might well be the most elaborate and rarefied mating ritual yet devised by humans, a form of sexual selection so vulnerable that if it ever attracts another mate with its song, the singer would already be extinct. The reproduction of our species survives as lost time propagated into the future. Sagan's "time capsule" sent as Druyan's "love letter" is an intergalactic mating call that has outgrown sexual attraction as a memorial for cohabitation. NASA's intimate relations with dolphins in the 1960s necessarily failed because humans behaved as the dominant species in the experiment, forcing cetaceans to speak on human terms.[28] This is why the long-distance relationship with the imagined "human+" aliens that Sagan posits for the Voyager mission is vital: in this scenario, humans are no longer in control, no longer so eloquent, no longer the dominant species.[29] We become the dolphins, as it were, and our shrieks and clicks on the Golden Record are the unintelligible sounds of our intelligence for another world to decipher as a call for intimacy.

Why should NASA's interstellar mating call be musical? Because music is an attractor, an overcommunicator, an interminable repeater, a mind trap. It overcompensates to overcome the gap. Human music, in particular, with its play of recursions — from rhythmic diminutions and octave doublings to formal expansions and motivic compressions — spins a web of intersecting cycles to catch the "ear" of another. What this ensures is that its recursive frequencies will somehow connect with the unknown, albeit inaccurately and inadequately.

But a mating call need only attract engagement. In our selection of nonnormative listening, it was not that nothing happened along the frequency spectrum; something happened . . . just not as we know it. Where an alien species may fold this spectrum is arbitrary and unpredictable, but the information music emits will attract a particular audition determined by an interface that will scale and process time in ways that we cannot imagine. And the other, too, if it manages to process the information, will imagine us in ways so strange that we will not recognize ourselves in their audition. And yet somehow, in this gap of miscommunication, music has connected us. The object may be opaque and withdraw itself from full disclosure, but it always mediates something of itself. After all, we humans already listen with estranged ears to bird "song" and whale "song"; their music attracts us in ways alien to their worlds. We have neither understood nor truly heard their song, but the music we imagine as theirs still connects us. Creative friction is the stuff of miscommunication.

Sagan was well aware of the communicative possibilities of miscommunication:

> Whatever the incomprehensibilities of the Voyager record, any alien ship that finds it will have another standard by which to judge us. Each Voyager is itself a message. In their exploratory intent, in the lofty ambition of their objectives, in their utter lack of intent to do harm, and the brilliance of their design and performance, these robots speak eloquently for us.[30]

In other words, Voyager cannot but be a semiosis machine, not so much a sign of meaning, but a sign of wonder, a marvel endowed with an eloquence that speaks of an otherness that need not be comprehended. It is an ineffable indexicality. It communicates as a shock to the system. It elicits attention. Whether it will mean what Sagan imagined is anyone's guess, but over time, a new network of perceptions, signs, and practices will develop around the object to fill in the blank. And who knows, given one of Sagan's "human+" aliens on a Goldilocks planet, an extraterrestrial recipient may be able to map out the recursive folds of our music by conducting some kind of statistical analysis, make the necessary "data correction" to reverse engineer the interfaces of transmission, calculate the temperature and density of Earth's atmosphere from the photos of clouds, dunes, and oceans on the B-side of the record, transform the information for its own environment, and work out from the level of rhythmic complexity of the music our degree of intelligence. Such intelligent life-forms would realize that it would be too late to reply to the mating call from Earth. But a reply was never the plan. We only need to get the aliens to fall in love with us, and given the music on the Golden Record, they very likely would.

Mission impossible accomplished.

INTERGALACTIC DOTS

Takahiro Kurashima

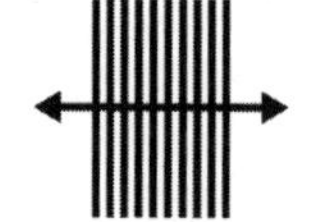

Use the film filter provided
with this book (or copy the
filter pattern in the Appendix
onto transparent film).

Coda

CHAPTER SIX

Definition

Musica est scientia bene modulandi.

—Augustine, *De musica*

A manifesto, a blueprint, and an archaeology of NASA's rudimentary music device merely provide a toolbox of blunt instruments. They are placards and sketchy strategies designed to arouse vision and stir knowledge into being. But their bold clarity lacks accuracy. Having paraded them, something more precise is required: a definition.

To de-*fine* music is to come to its end: *fine* (the musical term for "the end").

In order to close, then, we must draw together the various strands trailing haphazardly from our "musings" on the Voyager mission to arrive at a *new* definition of music. After all, the purpose behind this exoplanetary exploration is to defamiliarize the current state of music theory in order to launch it back into the universe without the outmoded metaphysics of Pythagoras. It theorizes what on earth humans were thinking when they sent music in the form of a Golden Record into a distant galaxy. In this context, we have listed seven premises for a theory of repetition and seven premises for alien communication, as well as a bewildering constellation of concepts, observations, and passing thoughts. By drawing these strands together, this definition will necessarily repeat what has already been said in the previous chapters: in this sense, it is not so much "*fine*" as "*da capo al fine*."

Closure, however, sounds too final. In fact, to close is precisely how not to begin a definition of what has been an open-ended speculation on music. Our intergalactic endeavor is supposed to challenge us to seek new discoveries across a boundless frontier and not to draw familiar lines. The attempt at closure, then, needs to be left ajar. A definition, in this context, should be less a repetition to end all repeats than a perpetual shuttling back and forth. In this regard, this definition is no cadence; it is a cadenza or embellishment that prolongs the end in order to generate new perspectives for further exploration. What follows, then, should not be read as a closure but, as dis-closure, a "*da capo senza fine*"—a repetition without end—intended to reveal, rather than to fix things in their continuous motion.

0001. DISCLOSURE

Music is a disclosure.

If music were not a disclosure, then there would be no need to *attend* to it. It would merely be evident, existing as some brute fact or unveiled totality that is apparent to all. Music would have no authority—no *reason* to be heard—because it would have nothing to disclose. It would just be raw immediacy. Pure presence. Totality. Music's mythic status and ancient theories tend to honor music with such attributes, as if music just *is*. But contrary to its fabled past, music is not some primal resonance, life force, or intervallic scaffold that orders the cosmos; it is neither the origin nor the essence of the universe. Rather, *music is embedded within the fabric of the universe as a point of disclosure.*

But what does music disclose?

Music is a disclosure of space-time. It is *an* expression of an absolute to which we have no direct access. Time is ungraspable; there is no scale or perspective that could account for its totality or unwind its functions. Music, however, by aestheticizing time and miming its qualities, discloses a particular way into its opaque operations. It opens a tiny window on time.

As a disclosure of time, music *comes in "piece."* It does not come in pieces, like the disordered totality of a jigsaw puzzle that might fall apart. Neither does it come as a bit, as if there is always something missing. Rather, as a piece, music brackets time as an entity where change is coherent. Music, then, is a piece of time. It discloses the *rhythmos* of a wider pattern of reality, but filtered and focused in a particular configuration, making time uniquely present. Music's disclosure is a temporal weaving that rhymes with the rhythm of a finely tuned universe as *this* piece at *this* time and in *this* place and as no other piece.

0010. EVENT

A piece of music is an *event*. As an event, a piece of music

- *takes* time and
- *takes* place.

It *takes* what are infinitely extendable properties — time and space — and makes them into discrete entities.

In taking time and taking place, the event *removes* time and space from their normal order, lifting them out of the ordinary and intensifying their qualities as an ephemeral flourish. It is like a node in a network that gives a point of audition where time suddenly snaps into place and then vanishes. Or it can be conceived as a weaving of space-time, a knot in the fabric of the universe that impedes its flow, drawing attention to its torsion before untwisting its patterns of coherence. As with Penelope's shroud, the weave focuses attention for a moment, buying time as it is made and undone.

An event happens. But an event is not *something* that happens in time; rather, *time* itself "happens" in the weave of music, caught as a texture, a pattern, a peculiar tangle of frequencies.[1] Time is "taken" in the event and reactivated aesthetically, undulating in the vocal folds of Tibetan throat singing, for example, or clattering through the wind chime of a bamboo forest, or warbling in the laughter of the kookaburra, or reverberating in a cavern dripping with rhythms.[2]

And in doing so, music also weaves its listeners into its folds, because by taking *this* time and *this* space, music coordinates a unique way of being in time and space. An event is the disclosure of our embeddedness in the temporal order. It is a dwelling.

OOII. FREQUENCY

Music's space-time event is measured in frequency.

Frequency measures time in three different ways: as a counting, as a relation, and as a metaphor.

1. Counting Frequency is simply repetition. It is rhythm. It should not to be equated with pitch (as the term is often used), although pitch is an integral part of the music's frequency spectrum. Music is measured in frequency, because music changes by repeating itself, creating the conditions for comparison in which identity and difference can be clocked and calculated. This can be measured on any scale, as long as the frequency can be counted.

As an interweaving of frequencies, music is a fabric of time where the interlocking of repetition and difference (A | V) results in a coherent texture that can twist and fold without breaking, generating further frequencies at recursive levels.[3] Music is multitemporal, but nonhierarchical. It is flat, because the different temporalities are woven, rather than structured, generating a potentially endless series of calculations that cannot be reduced to a simple formula.

Frequency measures music in numbers. These numbers are not the Pythagorean ratios of old, suspended as proportions without sequence or temporality. These are not abstract, silent numbers. Frequency is literally a measure, an interval that counts out loud, marking each rotation as it rearranges the order it inhabits. By counting out loud, frequency is not some outside system applied to music, as if it were a form of analysis or mathematical modeling. *The measure is the music.* As Saint Augustine famously remarked, "Music is the science of measuring well" (*Musica est scientia bene modulandi*).[4] It is an arithmetic of rhythms, a numbering that maps and scales the quality of time. In this sense, an event frames an intersection of infinitely repetitive cycles into a particular pattern, assemblage, or network of numbers. The ear hears numbers because music counts these cycles out loud. And since our ears are basically counting machines, when we listen to music, we are secretly auditors and accountants.

As a counting object, a piece of music exists "somewhere" along the frequency spectrum; its exact location depends on how it is "audited." In itself, the frequency spectrum is nothing more than a flat, undifferentiated terrain of numbers ordered sequentially into infinity. This glissandolike sheet, however, is shaped in accordance with the receptive particularities of each species. Music is always made to measure. Different physiological receptors will result in different computational solutions that crease the frequency spectrum at various junctures of perception, folding the sheet into a piece of origami with various surfaces for timbre, pitch, rhythm, and form. Ears (or whatever "counting machines" an alien species may evolve) measure these complex modalities of repetition. From the ears' "perspective," the flat surface has recursive folds that enable the perception of different rotational scales. Ears, then, fold music into being as origami formations that, depending on the species, will be of vastly different sizes. These musical structures are holographic in that their three-dimensionality does not exist except as perception.

Music takes place in the differential space between frequency transmission and frequency reception; what "appears" in this space is not so much a formation as a transformation caused by the difference between transmission and reception. Their contact is the interface for the disclosure of the event. Without the interface, music cannot happen, and yet the interface is music's most precarious juncture precisely because of the difference between the two sides; they can touch each other only indirectly. This holds true for all interfaces in music's life cycle—between beings, objects, or a mixture of media. The contact point is therefore always a point of uncertainty that remains vaguely opaque. In this sense, all contact is alien contact and demands the mobilization of counting machines on both sides to connect through complex computations to prevent a face-off at the interface. Miscalculate the "differential equation," and the frequencies will fuse and short-circuit — nothing will happen. Or more likely, something unexpectedly odd will happen that forecloses the disclosure — the event becomes a nonevent. But sometimes alien contact gets lucky and results in an accident of "*creative* accounting" — and new music happens.

Music, then, cannot be defined apart from its mediation. Music is a material flow through different surfaces, a mixed compound of frequency modulations riven by various interfaces. In the "science of measuring well," the medium is not only the message, it is also its measure.

2. Relation Musical events do not simply *take* time and *take* place, as if time and space were substances which the event subtracts, seizes, or consumes as its possessions. Music is a predicate of time and space — it times, it spaces, it measures, it events.[5] As timing and spacing, an event is not merely a *taking*, but a *making* that *gives* time and *makes* space. Time's aesthetic dimension in music is a creative act, a donation — quite literally a "present" — that particularizes time as a gift to be received. If music offers a *unique* rhythm of inhabiting or coordinating the universe, then its reception cannot simply be defined in terms of frequency: it is a frequenting, a repeated form of visiting or attending, a habit that donates *this* time and makes *this* space as a suggestive habitat. Music's disclosure is an opening for an-other to frequent its frequency. It gives time so that the other can keep time. Without the possibility of this relation, there is no music.

Frequency, then, is not only the measure of music but its very *relation*. It is a counting that accounts for an-other. Or to put the matter the other way round, if music "is a hidden arithmetic exercise of the soul," as Leibniz claimed, then it can be so only as a disclosure of frequency that gives time for another to count.[6] Counting is a substrate of knowing, an exercise in measurement that learns by repetition, converting motion to emotion, patterns to memory, vibrations to sympathy, and resonances into imagined universes. So although music is frequency, it cannot be *reduced* to frequency, as if it were some containable substance, because frequency is precisely what is irreducible; it proliferates. And, since everything in the universe repeats, it is the connective medium that forms the relation between things. Far from being reductive, frequency is a generative and integrative process, spinning patterns of recognition and weaving knowledge through feedback loops and reiterative sequences. The universe learns by rote, and music brings this rotation to audition, if not to addiction, as it teaches its listeners to *keep* time and its practitioners to *make* time. Frequency is a site of inexhaustible creative collaboration.

3. Metaphor But how exactly does frequency make time for the other? Music aestheticizes time by extracting its qualities and reveals time obliquely as a metaphor of frequencies. To frequent time through music is not a direct mapping of time, but a metaphorical transformation that remakes time. The metaphor "time *is* music" does not mean that music is a subset of time or that music contains time; this would merely be a reversal of the metaphor (that is, "music is time"), ascribing the property of time to music, which is both banal and obvious. Rather "time is music" equates a melody, a groove, a beat, or a dance with time itself, as if two *different* things are one. The metaphor fuses an opaque, ungraspable object (time) with a sensual entity that is different, yet obliquely related (music). According to Graham Harman, the only reactor where this fusion can take place is the one that receives the music. The auditor replaces the object by enacting the metaphor; to frequent music, then, is to be in time with the metaphor. In fact, we humans become time in music. Take the curvature of space-time in Einstein's theory of relativity; this is of a dimension that is inaccessible to human experience, but a piece of music can shape qualities of that curve and fold us into its temporal space; we can follow the curve and even enact its timing. In experiencing music as a metaphor of frequencies, we become time incarnate.

Frequency, therefore, is a metaphor of time. This metaphor should not be misunderstood as a literary device operating virtually in the imagination. It is a physical play of frequencies, a resonance that literally moves whatever body lies within its bandwidth. Frequencies *matter.* They are not metaphysical numbers, but physical metaphors that vibrate in and through and between things; they initiate contact and are not dependent on some agential consciousness to conjure them up.[7] Whatever tunes into their vibrations counts, and through their metaphorical materiality, different objects (animate or inanimate) resonate in rhythm. Although each object would receive the frequencies differently, depending on its particular constitution, the metaphor is communal because it is physical, opening up the possibility of an "interobjective" timing between humans, between species, between objects.[8]

Thus, as metaphor, frequency is always a potential ontology for a relation — a hospitality of time — in which different entities can count aloud together in rhythmic counterpoint. Frequency as a counting, as a relation, and as a metaphor may be different ways of defining music, but they are ultimately interwoven.

0100. INFORMATION

Music is information.

To keep time is to store frequency as data. In storage, the musical event solidifies; its temporal *form*ation compresses as *in-form*-ation, internalizing time as space in order to preserve its presence. "Inform" is the form of music effaced in storage. It is the potential of data for communication.

Data storage is repetition in waiting. It is implied in every event, because music is not only an *ex*-pression of its embeddedness in the universe, it leaves an *im*-pression, as if it wants to be remembered. If music were merely expression, then its data would be lost the moment it passes. It would be an ecstatic, but ephemeral flicker. Music, however, repeats itself compulsively. With its constant rhythmic reiterations, music carves a signature, placing an imprint on the objects that encounter its fleeting appearance. Even when it disappears in the fabric of space-time, music always leaves a trace, because repetition is the very substance of memory and recognition; it lingers as the stuff of identity, as a distinct mode of persisting, etching its cycles on the materials it touches, from the neuroplastic to the ferromagnetic. Like the bends in a knotted string that has been unknotted, music can be recalled and can easily reknot itself. As such, music is not only an internal process of repetition, but is *inherently repeatable*. Playback is built into its being; it is music's reproductive cycle, as it were, engendering an-other event of itself. Playback enables music to reevent itself. A "piece" is therefore the possibility of a recursion at the highest level, the repetition that ends in storage in order that the event can recur. Data storage should not be regarded as a technology apart from music; rather, it is part of music's life cycle. But this technology should not be mistaken for human-made machinery. Listening, for example, is already the potential for playback. It is an act of storage in real time.

Music, then, is a binary object, a conjunction of event and storage in which sounds glimmer in time, then dim into space. Music blinks on and off. This musical on-off switch has a long history. Augustine in *De musica*, for example, distinguishes between music as space, structured by the Pythagorean ratios as a silent order outside time, and music as movement, which is the rhythmic horizontalization of the cosmic ratios. The two forms coexist as the well-measured temporalization of well-tempered proportions. In updated terms, the event is music as movement, and data storage is akin to Augustine's music as space, although they do not so much coexist as alternate along the same time line. As with Pythagorean numbers, once stored, the frequencies can no longer move; they curl up in space, waiting to unreel themselves in time. But unlike their metaphysical counterparts, which are detached from matter, in data storage, music's counting is a count-*thing*; its transformation from verb to noun is a materialization in which time is locked in space and transported in silence out of its own time in order that it can be played back anywhere and in any future time, not as an eternal number that is everywhere, but as a particular peripatetic object.

Music's information technology, then, converts time into space, creating a *piece* of time that can be transported into another time and into another place. Because music is always discretely *this* piece at *this* time and in *this* place, its recurrence is always out of time and dislocated. Long before Voyager's mission, music was already a spacefaring and time-traveling vessel, carried from human to human in aural cultures, through notation in written cultures, along the pits and grooves of analog cultures, and as binary bits in digital cultures. These hosts are all prosthetic extensions of music's compulsion to repeat, enabling it to blink on and off as it circulates and disseminates across time and space.[9]

So when music is "switched off" in storage, data is static in time, while the machine can move in space. Conversely, in an event, the machine is stationary, while data moves in time.

Music is sometimes cut in half, defined as either "on" or "off." It is, of course, both (but not at the same time), and neither half should be privileged above the other. There is no need, for example, to choose between a score and its performance. Music simply changes state. Each state is radically different.

1. In event mode, the action is switched "on"; music is not so much "music" as "musicking," thriving on real-time agencies that require a unidirectional accounting of change. It is unpredictable. The event is always local and ephemeral, because its *location is fixed in order for time to move*. This is the site of performance.
2. In data-storage mode, time is suspended in space, and music assumes a thinglike materiality. Information disseminates and persists, because *time is fixed in order for location to move*. Music's repeatability is the site of analysis. This is because the conversion of time to space—whether in the memory, on paper, on vinyl, or in bits—petrifies motion, enabling the analyst to count and recount music in any direction and at any speed. Music's

repeatability is repeatable in any order and in any time, until it is reactivated as another unidirectional event. Data storage is the site of pre-formance.

Both modes — action and reflection — are necessary. Their relation can be theorized as a musical metastructure: just as the tiniest repetition in music is a fold (A | ∀), its largest repeatable unit (the piece) creases back upon itself as an event/storage mirror image, with the interface functioning as the third element around which everything hinges. As the mediator of difference between transmission and reception, the interface constantly registers change, filtering the information from one state to another as the music folds back and forth like a concertina in a series of shape-shifting variations (A | ∀ | A | ∀ and so on). The *pre*-form of storage will never actually result in the *per*-form of the event. Each re-event will introduce new contingencies in time, and the material peculiarities of each medium will alter the musical substance. Repetition is always a disclosure of something new, even if it seems to repeat the past. While this long, unfolding series would be perfect as the stuff of reception history and media archaeology, in practice, one fold is sufficient to define music.

Music's binary life cycle can be summarized in the following diagram, Figure 6.1, which should ideally repeat the sequence across several pages in different "sizes" to register change across time and media.

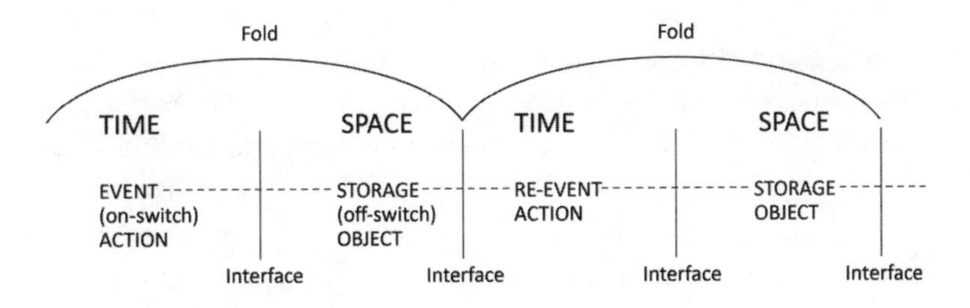

Figure 6.1. Music's binary life cycle.

OIOI. NETWORKS

Music is a network. It networks in two ways.

1. *Internally:* music is a weave of frequencies that entangles time as a network of relations. This is a self-rotating network. There are no strings attached from the outside to manipulate its inner workings, and there are no loose ends dangling from its boundaries to snag its autonomy. This is music's innerface.
2. *Externally:* music is embedded in an interconnected network of mediations. At a minimum, it is simply the reaction of receptors making sense of frequency information. For the "human+" aliens envisaged by Voyager's team, this information expands into a giant, generative force: music is caught in a web of meaning making, spinning itself outward, even as it is being spun by external forces in a swirl of semiosis.[10] Music interacts dynamically with a network of discourses and objects that coordinates its external relations and constructs its cultural meaning. This is music's interface.

The two networks coexist independently and are indirectly related. Their overlapping autonomy is analogous to the function of Penelope's shroud. On the one hand, music is a holding pattern with its own temporal integrity, woven and disassembled with a "purposiveness without purpose" that remains opaque to the external network suitors;[11] its rhythm simply appears and disappears, blinking day and night as an autonomous activity shorn of any use except to keep time—it waits, it counts, it withdraws. On the other hand, music is simultaneously a shroud of intrigue and incipient violence that casts its net across a narrative of Homeric proportions; its meaning is entangled in cultural practices, moral dilemmas, social relations, gender roles, power plays, and so on. The same weave is both texture and text. Both belong to the object. Texture is its structure, and text is the boundless possibilities of meaning emanating from structure. One is in the realm of theory, the other in the realms of media. Together, they loop as a Möbius strip.

Although the two networks coexist, their relation is inextricably bound to the binary structure of music as event and storage. The rate at which music blinks determines the validity of each network:

1. The quicker it blinks, the wider the dissemination of the external network.
2. The slower it blinks, the greater the density of its internal network.

When music is switched off, it becomes a "dark object" that travels in time. It is "windowless," both monadic and nomadic, disconnected from the external network that locates its meaning in a particular time and place. Because music time travels every time it is switched off, each time it flicks back on as a "bright object," music has to reconfigure its external networks in a different time.[12] How much remains of the older network depends on how long the switch remains "off." A piece of K-pop blinking rapidly is likely to light up its original external networks while continually amassing new ones, creating the possibility of endless heterogeneous mediations; its event would be a brilliant mesh of cultural specificity that sprawls outward, with tendrils of meaning glowing far into the distance. Such networks coagulate into genres, coalesce into symbols, merge into objects, divide into new categories, assemble into practices, and solidify into institutions as the light pulses back and forth. It empowers music with representation and endows it with language; it mediates social agency and propagates its meaning. Storage allows for such dissemination. But the further music travels *in the dark*, the weaker these external networks become, eventually losing all ties to the wider environment of meaning making in which the piece was originally located; the historical and cultural specificity is lost as the network of relations disintegrates, leaving nothing but a dark object of data. Music's distributed being withdraws into dark matter.

Most music dies in the dark and fossilizes as remnants of data storage. For pieces that suddenly flicker after years of disuse, new networks may arise, but because the network around each re-event increases in entropy the longer it is dormant, the interface will eventually collapse into a chaotic and arbitrary tangle of relations. Voyager's time capsule is a symbol of the longest "off switch" in music history: if it is ever switched on, our music will be lost in a distant time and an alien territory. What is certain under such conditions is the principle of uncertainty. Music's uncertainty principle has two laws.

1. The longer music is stored, the less precise the external network; the external network becomes increasingly unpredictable.
2. The less precise the external network, the more precise the internal network; the internal network becomes increasingly determined.

The Golden Record is a limit case. If this object falls into the tentacles of an alien life-form, all that will remain of the music is frequency: loops, cycles, folds, oscillations. The texts and contexts that defined its shape will dissolve into a mass of texture. Music will simply weave time. Its fabric will be shorn of the extraneous connections that had constructed its meaning. Indeed, everything on NASA's shiny disc will be stripped of meaning and become music; the murmurs of human speech and the sounds of nature will just be a play of frequencies, and the bands on the LP, like the rings of Saturn, will just look like a binary rhythm. If another intelligent life-form replays the Golden Record, the decontextualized event would be the site of an alien structuralism. Earth becomes absolute music.

0110. ESTRANGEMENT

Music is an alien form. It estranges.

In a pro-alien music theory, the first step is to acknowledge that music is not solely made by human hands, as if it were already in our grasp; it is not even for our ears only, as if we have a monopoly on its frequencies. Music predates *Homo sapiens* and will outlive the species. It is found in things; it sounds in other creatures; and if the theory of evolutionary convergence is correct, it is likely to echo in other species in faraway galaxies.[13] Music estranges. In fact, human music already survives in this vein, blinking on and off as peripatetic time machines, awaiting strange encounters in other regions and different epochs. When music's dying strains reach their final outpost as a shiny disc on a distant planet, its alienation will be complete. And yet it is precisely in this moment of total alienation that music reveals itself as an unalienable possession that cannot simply be consumed or exchanged; its sheer existence comes in excess, as a gift without return, not so much to be given away, but to be passed on as the inheritance of an entire species to another for the safekeeping of its time.[14] Unalienable possessions are ultimately for aliens.

Penelope's weave is an unalienable possession. In Hellenic culture, women wove to be remembered. Their fabrics were given as parting gifts. These garments were an emblem of hospitality, a form of portable housing for the wayfarer (or in the case of the funeral shroud, a vessel to accompany the dead); their very texture and pattern bore the signature of the giver, as if the weaver had woven herself into the very fabric that clothed another's body.[15] Textiles authenticated a unique relation between strangers. When Odysseus returned to Ithaca, he was a stranger; he was someone whom Penelope already knew, but could not recognize, and it was by confirming the garments that she had made for him that their estranged identities reconnected.

Something similar stirs in Sagan's vision for Voyager: for him, aliens are strangers that we somehow already know, because we believe the music that bears our signature will resonate with them and trigger an estranged recognition of a garment woven in our time and our place, but now radically untimed and displaced. We weave music to be remembered. We give music as an act of hospitality, as a fabric of time to clothe the stranger. If aliens were to recognize a shared intelligence through our music on the Golden Record, then our estrangement will become their keepsake, and our space odyssey will arrive home as a memorial — a funeral shroud — for our species.

We won't be around to witness the event, but perhaps we should imagine ourselves on that distant planet now and assume an alien form, because all encounters with music involve a double estrangement: not only does music estrange time in disclosing its peculiar qualities as a metaphor of frequencies, but music is itself estranged in time and dis-located in space. While we might imagine this repositioning as a celebration of diversity across the Earth like NASA's mixtape of World Music, what is alien is not, in fact, the cultural difference coordinated by the external networks. Strangely, what is alien is found precisely in what we share in music—the fabric that makes difference possible. As music estranges itself through time and across space, shedding in its wake a series of crumpled networks that no longer mean anything, what survives is the "in-form" of frequencies. This inner network, drifting in the blackness of space, makes for a disorienting encounter, and yet this disclosure is precisely what is shared across cultures, species, and galaxies. It is unalienable. What estranges us in music, then, is the *disclosure* of an ontology that connects us—a way of being in time.

When music is received as an alien structure, it invites its recipients to enter a strange dance with its vibrations. Its rhythms do not only indicate that there is intelligent life elsewhere in the universe, but that the universe is itself intelligent. The wonder of music is that music can exist at all, that time itself can be wonderful, inviting us into not only a measured existence, but a well-measured one. In this sense, Augustine is correct to hear music as an opening into the "hymn of the universe"—a window onto an ecstatic cosmic order.[16] Or in the terms of NASA's pacifist liturgy, music "comes in peace": music is the piece that makes peace possible across galaxies. In both belief systems, music ac-*counts* for the ultimate as a particular and peculiar measure; it is a science of measurement for disclosing what is totally immeasurable. In the end, whether an alien hears the cultural difference between a Navajo chant and a Bach fugue in the needle on the Golden Record is a moot point; it is all frequency—just one Earth, just one universe. Of course, there will be frequency differences between the two pieces, but *what counts is time*. NASA has sent a time capsule into space; it literally shares time, or more accurately, it shares miraculous *pieces* of time found on a little blue planet, curated in the hope that a stranger in the distant future might one day keep time with us.

★ ★ ★

Here ends the six-fold definition of music — disclosure, event, frequency, information, networks, and estrangement — although, given its compression, it should probably be repeated at half speed so that it can unfold more slowly: *da capo al fine*. This definition is pleated, because music itself folds in and out, not just in its disclosure of time, but as something in time. It has a life cycle. It metamorphoses. It re-events itself in different states and experiments with alternative realities. Music makes and sheds meaning. It glows and dims in the dark, inverting its temporal and spatial dimensions. It shape shifts across interfaces. In IMTE, music is time emitted as an item across an ever-changing landscape, rotating like an anagram that rearranges its elements to adapt to the terrain. As such, music's definition is less something to fix than something to track. It might even go off the radar. Sometimes it is only at the point of disappearance that objects lose their familiarity. Defining music in the terms of Voyager's mission entails thinking on a scale where we might lose track of our grasp on music. But if we are prepared to lose it, we might just find it. And finding that place might be as far away as the ends of the universe. To define music is to come to its end.

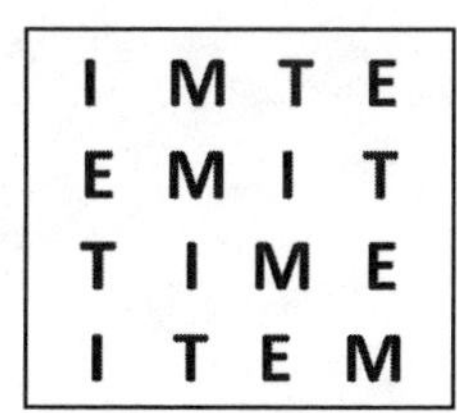

CHAPTER SEVEN

Repeat

What always matters is folding, unfolding, refolding.
—Gilles Deleuze, *The Fold: Leibniz and the Baroque*

The end.

Finally, our space odyssey in this TARDIS-like book has arrived at a destination. When you venture out, music should be different. It should be in a different time and in an alien space. We started out with a few blinks—just simple dots on a page—and ended up with a densely woven definition, pleated in a complex pattern of repetition and difference. It is as if the initial blinks were just the pinholes of a needle poking through the fabric of space, threading a definition of music under the propulsion of the Voyager mission. But our weave is just the beginning. There are still so many loose ends. And they are deliberately loose, because by coming to a frayed ending, we are handing the mission over to you. This is not for you to stitch up, of course, but to continue the warp and weft in different directions. So our final chapter is not a conclusion; it is a briefing that summarizes our strategies for you to pursue an intergalactic music theory of everything.

"What is music?" is the persistent question tugging the pages of this book. Eschewing the normal models, our definition has been an emergent process, assembled from bits of space debris as we journeyed across our varied chapters. An emergent definition means that music always takes flight into the unknown. The unknown, of course, is space. By positioning music in this interstellar context, we have attempted to decontextualize the terraforming of music that habitually claims to know music's being. The alienating perspective enables us to speculate on what music might be when it simply exists. Music *is*. And it is precisely because music *is* — without the need for human interpretant — that contact with music is always a serendipitous and fragile enterprise that should not be assumed. This book, then, can only be a blueprint with a number of tentative coordinates to plot lines of intersections.

- 1 question — What is music?
- 3 rules — simplicity, inclusion, and embeddedness — plus
- 3 parallel concepts for an intergalactic theory of everything;
- 7 premises for a theory of repetition.
- 7 premises for alien communication, and a
- 6-fold definition of music.

All in all, there are 27 points with which to chart various emergent paths across the Milky Way. Please do not arrange them into a cube, as if music were a solid object to be grasped (3 × 3 × 3). Scatter them as a constellation to navigate the rhythms of the universe.

Given this immeasurable context, the question "What is music?" is a kind of limit (or a measure) that inverts the infinite possibilities of space-time into a particular point of view — a piece of time. The fact that NASA sent music into space as the main channel of communication to alien life-forms indicates that there is something exceptional about music, some kind of knowledge that is possibly more primitive, more explicit, more universal than the science that rocketed the mission into space and the concepts that argued for its feasibility. So even as we launch music into the unknown to estrange its own definition, its echoes promise a sounding of the beyond. Of course, the difficulty in attending to this sonic mirror is to avoid catching ourselves at the center of the reflection. It is almost impossible to be consistent in banishing ourselves to the margins. But if, even for a moment, we can squint sideways in exploring what music is, then we might just register something that is potentially an echolocation of everything.

In current thought, there are two competing "theories of everything" that claim that title. It is a tussle between science and the humanities.[1] String theory, which aims to unify everything in its search for quantum gravity, stands in one corner (if it can find its way through its eleven dimensions); in the other corner stands object-oriented ontology, which also goes by the snazzy name of "OOO"; "triple O" (as it is pronounced) claims to be a theory of everything by incorporating every*thing*, relating objects of any size and complexity, concrete and fictional, in all their opaque diversity and autonomy. So there are two theories. One deals with strings, the other with things. IMTE, of course, is not offering itself as a third contender, but it does straddle both, with one foot entangled in strings and the other firmly planted on things. This does not mean that it wants to unify the two theories or endorse them. In fact, it barely scratches their surfaces and merely deploys bits of both to navigate the universe of music with NASA's little space probe as method. Their approach to everything just happens to fit our enterprise. And in the spirit of alien interception, we have used our various tentacles to collate other theoretical debris orbiting the two contenders. To picture our exomusicological enterprise, it is best to imagine a space-faring craft with two magnetic antennae protruding from its tail — *media* and *theoria* — each attracting whatever falls into its field of influence. Conglomerating around one antenna we have various flattish-foldable-stringy-dotty materials, and on the other, we have accumulated a clutter of object-oriented-speculative-posthuman-networky things.

IMTE *Enterprise*

Maybe all this looks a bit clunky and sounds somewhat awkward. But then, what we have cobbled together merely imitates the Golden Record and NASA's makeshift gramophone. It's an accidental homage. As with Carl Sagan's team, we have curated a sample of what's out there without much context, mixing and matching incompatible theoretical genres in the hope of transmitting our ideas through a basic scholarly medium to reach the widest possible audience. Each item in our cluster of curiosities is designed to spawn its own response. It is about increasing the potential for alien contact. It has been a wonderfully indiscriminate process, meandering across a flat ontology devoid of hierarchical divides where any music, animate or inanimate, and any theory can lie side by side like tracks on an LP. Having one foot tangled in strings and the other on things is just a way to organize our intellectual kleptomania. This is not because we are too undecided to trust a single theory. Rather, it is because music comes first as the object of theorization. The cluster of theories assembled around the object are secondary resonators that vibrate sympathetically to its frequencies.

Of course, we could have just targeted our native crowd of musicologists, which would have required *no* foot in either camp, but by taking two small steps for music, we've tried to make a giant leap for musicology. Musicology — now with "exo" as its neon-lit prefix — is planting its flag on "everything." It wants a stake beyond its narrowly defined enclosures and to communicate across disciplines as something that is not only relevant, but transformative in the sciences and humanities. Of course, planting a flag on everything should not be taken as an attempt to colonize everywhere; as with NASA's naive pacifism, our work is a donation, flagging music as a communicative vehicle across frontiers. After all, there is no point waiting a few billion years for a reciprocal act of cultural diplomacy from another galaxy. We should simply start now, with IMTE setting in motion a gift economy circulating across various disciplinary nodes dotted around our planet.

We have attempted to redefine music for everyone — simply and without jargon. With sufficient delay and the difference necessary in the etiquette of gift exchange, we hope to see a host of incoming modules zooming back at warp speed and locking together to create an ever-generative theory. Taking heed of the slim chances of success for Voyager, we have tried to boost our chances of contact by sending many gifts, rather than one golden payload, scattering them across our text in the hope that someone out there in the darkness might intercept our peace offering. These are packaged not simply as arguments and propositions, but come in all shapes and sizes — as rules, axioms, acrostics, puns, aphorisms, cartoons, diagrams, blueprints, manifestos, anagrams, long lists of things, moiré patterns, and the occasional opaque turn of phrase. So if you are serious about hitching a ride with us, comb the text for gifts that might just open out like tiny wormholes into an unexpected universe for your thinking.

Our claim is that music matters everywhere. It may seem that we are inviting you to get into the groove and take a spin on a golden disc, allowing music to weave its rhythmic loops around your thoughts so that it might string us together in higher forms of communication. There may be a sense that we are advocating music as a universal language through which our different disciplines can be translated and understood. This is a nice idea, but it would also be a terrible mistake. Music does matter everywhere, but it is not the answer to everything. It is a question. In Douglas Adams's *The Hitchhiker's Guide to the Galaxy*, the oddest thing in the universe is a small yellow leechlike creature known as the Babel fish that feeds on the brainwaves of others and excretes a telepathic matrix of frequencies into the host's brain. It is the equivalent of a Rosetta stone for intergalactic communication. You simply stick the fish in your ear and instantly understand any form of language received anywhere in the universe.

This decoding organism is the answer. It is the elusive interface to mediate all difference; it is the bridge that cannot be built into communication technology because the unknown other is always a "blank" particularity that short-circuits the system with its own protocols; it is the impossible solution between the Golden Record and its alien receptors, between the vibrating stylus and the "whatever" out there that transforms, misshapes, undoes, repurposes, or ignores the signals. The Babel fish is the missing link that is always going to be a gap and therefore cannot exist; hence it is necessarily a fiction. In *The Hitchhiker's Guide to the Galaxy*, its existence is so improbable that it proves the existence of God, except that in a twist of theodicy, the unfortunate fish, "by effectively removing any barriers to communication, has caused more and bloodier wars than anything else in the history of creation," rendering the proof meaningless.[2]

As the eraser of difference, this dream interface is not only improbable, but precisely what not to wish for. It's the proof that destroys itself. The Babel fish is what we must let go, let be and not know if there is to be the contingency and the element of surprise that are vital to a genuine gift economy. We need a technology that preserves the nonidentity of the other and a music that communicates as difference, for they point to the ontology of peace that NASA's mission presupposes. Why else would Sagan's team send the Golden Record into the unknown as a love letter from humankind? In contrast, the removal of all barriers is the precursor of violence and the technology of power. We need the gap, because the gap in the system is the synaptic possibility for connections that can make a *difference*.

Music, then, is not the "universal language" that translates everything, but the universal gap promised by data storage for the synaptic moment. Music as an event is therefore an opening; it is a question, a blink that draws attention to a gap. And it works both ways. On our end, it is configured in NASA's sound technology quite literally as a gap with the omission of speakers. On the other side, the extraterrestrial also has to make the leap; it also enters a blank where there is nothing to know. It must delve into a sliding scale of recursions that, although shared between us, in fact calculates our differences depending on what each side brings to its measure. There are merely strange vibrations in this connection, opening a pattern of communication along this frequency spectrum that is utterly alien to both parties.

Who knows what music means? It may be that Voyager stumbles upon a small, ephemeral life-form whose life span is precisely six minutes and thirty-seven seconds. Having exhausted several generations to construct the gramophone, these aliens discover to their astonishment the final track on the Golden Record. In each generation to come, one from their number is chosen to experience the most exquisitely painful existence possible, hanging upside down as the cartridge stylus vibrates to the rotations of Beethoven's Cavatina: the music pulsates through this chosen body like a throbbing dance that structures its life in a ternary form so fulfilling that at the very end of six minutes and thirty-seven seconds, its being expires in a state of unimaginable peace. No human being could experience the Cavatina with such intensity of meaning, not even Ann Druyan and Carl Sagan, for whom the Cavatina was "their song." On the other hand, it is equally possible that the final track of the Golden Record is perceived by some alien behemoth as nothing more than a sharp, pertinent fart, no different from an accidental audition of the same Cavatina as an image on the B-side of the record.

Such is the banality of the gap. There is a strange ethics in communications technology, where the belief in the possibility of miscommunication is the key for our rhythms to cross in creative and productive ways — or not. So technology is not designed to "solve" communication, rather it is its potential failure that makes communication possible. Voyager is the gap that bridges the unknown across the universe.

★ ★ ★

Meanwhile, to return to this TARDIS-like object you have been reading here on Earth, the same principle applies. Its pages are designed to unfold as a platform for different things to connect. It is an interface through which the data stored between the covers are released as multiple contact points. But this book is no Babel fish for interdisciplinary transparency. While it has pretensions to connect everything through music, it does so as a skeletal framework, creaking with missing bits that disconnect us. It is a blueprint. A manifesto. An initial archaeological dig. It is a gap, akin to NASA's missing speakers, for we have deliberately "failed" to apply the theory and to demonstrate its potential in case an example becomes exemplary, authorizing an official way of application that would narrow communication and shut down creative encounters. This book is less something that works than something that "wants to believe." This is not some vague hope in an unknown "X" out there, but the need to entrust ourselves to a music that presupposes an ontology of peace in which reciprocation is possible. It is on good faith that this gift, "dispatched without hope of return,"[3] as Timothy Ferris, the producer of the interstellar record, describes the Voyager mission, is given as a true gift. Somehow, the universe will make good on its promise.

Without positing such a universe first, there really is no point theorizing music in space. Ontology has to come before human technology. Prior to any mission is the belief that music already is, long before *Homo sapiens* inhabited this planet — in resonant caves dripping with rhythms, in the accidental counterpoint of creatures in the canopy of the forests, in the sea and under the sea in the thermal gurgling of volcanic vents, in a music that etches itself on things in time through space swirling in billions of galaxies, in which we, this strange loopy species on Earth, have found a way of being and a way of giving.

Music blinks because it sees us first. Its rhythm opens a gap for reciprocation. In this sense, we don't make music; it makes us. Why do we employ all our technical know-how to send music in space? Perhaps because we know intuitively that it is a gift that we have received prior to our being, and we are simply reciprocating, creating a rhythm of donation that might ripple across the universe. After all, by sending what President Carter called "a present from a small, distant world,"[4] the Voyager mission has set in motion a gift economy for "all worlds, all times."[5] This principle can operate on any scale, from the village to the universe; it is just that Voyager's cosmic vision magnifies the ethics of alien hospitality to the point of ineffability. The Golden Record is cultural diplomacy writ large from the good Earth. And the task of exomusicology is to formulate the conditions for NASA's symbolic act to circulate as humanity's peace envoy to the cosmos.

For some skeptics, the belief in universal peace is not only naive, but dangerous. Very dangerous. Not only has NASA wasted its resources on something as useless as music, but the space agency has given away the secret of Earth's location, first on the Pioneer 10 mission and then again on the Golden Record, with a map showing the coordinates of our sun's nearest pulsars; this is tantamount to an open invitation for an alien invasion.[6] There is no universal harmony in a star-wars ontology, only a "dark forest" teeming with suspicion and cloaked in silence.[7] A universe predicated on mass destruction is without music and devoid of the possibility of communication.[8] No one can listen to an-other: intelligent civilizations are either about to attack or hiding in fear — and earthlings, with their little space toys parading in orbit, are just stupid. In such a world, Voyager's glistening disc of peace would just be pie in the sky, and our attempt at an exomusicology would merely be an obtuse exercise, rendering this project on species counterpoint pointless for our species.

A star-wars invasion is not the only inconvenience that would scupper our belief in the Golden Record and the vision of exomusicology. All we need is a tragic universe. In *The Birth of Tragedy*, Nietzsche imagined cosmic reality as a metaphysical loom with Wagner's music dramas as its inward illumination; it is "as if we were seeing the fabric on the loom while the shuttle moves back and forth," he writes. In other words, music represents both the appearance of an order and the tragic reality behind the illusion. Or in Nietzschean terms, the diaphanous fabric is the veil of Apollo, whereas the shuttle is the mechanism of Dionysian indifference.[9] This oscillation of deception and truth could also describe the music on Voyager: it is beautifully tragic. After all, a common answer to the Fermi paradox is one of Dionysian destruction: given the multitude of Earth-like planets in the vast array of galaxies that are billions of years older than our young solar system, there must be intelligent beings out there who can find us — so, where are they? One answer is that they are not here because intelligent civilizations destroy themselves. Intergalactic communication will always be too late. The *L* in the Drake equation — longevity — is the barrier. If *L* is short, then, as both Drake and Sagan imagined, it is because advanced civilizations simply blow themselves up. *Ka-boom!*

This answer, of course, is a projection of a Cold War ontology onto the alien other: the tragedy is not out there — it is human intelligence that is bent on total destruction. Our ever-expanding gains are unsustainable, and we will eventually self-implode in some nuclear holocaust or (in contemporary terms) through an existential crisis of the Anthropocene.[10] In this scenario, Voyager's time capsule, which was deliberately sanitized of all references to war, disaster, and disease, carries a lie about our species that, like the veil of Apollo, masks the truth of our future self-destruction. The beautiful music would merely accompany the fateful machinery of our own annihilation, like the final act of Wagner's *Ring*.[11]

But then, Sagan's team would never have sent music into space if they believed in this answer to the Fermi paradox, since no intelligent life would survive long enough to connect with another. A tragic universe does not need the gift of music to weave an interstellar handshake between species. It does not need to keep time with difference. Voyager's mission makes sense only if we suppose an ontology of peace that music always promises. "We step out of our solar system into the universe seeking only peace and friendship . . . and it is with humility and hope that we take this step."[12]

Attributing peace to the universe (and its musical soundtrack) may indeed be a naive mistake: it is an abstraction that is as hopelessly Romantic as it is without proof. However, it is precisely because peace is not universal that peace can exist. Peace is always particular. It is a relation. It is not something that is, but something that comes. By coming in peace, NASA's minuscule probe turns out to be a universal sign, because the gift it carries is the only known object outside our solar system that promises to bring peace across the universe. If this musical offering is ever understood by another intelligent life-form, then the Golden Record will be the ontological proof of the existence of universal peace.

The gift is the answer to our question. Like Penelope, Voyager keeps faith with an uncertain universe by weaving music as a holding pattern for the future. In ancient Greece, the loom was often thought of as a mechanical lyre, providing the underlying rhythm for a weaver to sing her song.[13] Perhaps, when contemplating Voyager's mission, we should imagine Penelope singing under her breath as she weaves and unweaves the fabric of time; she is whistling in the dark, spinning a tune as if it were a thread of hope that keeps the chaos around her at bay.[14] But her rhythm is not merely the drumming up of courage in the face of despair. Penelope's weaving is a fabrication that deceives and ultimately condemns her suitors to Hades.[15] Her craft, in its seeming innocence, is a crafty act of computation that ensnares her detractors in its web.[16] *Techne* is a form of deception. It is an art. And who knows? Voyager's naive *techne* may itself be a serendipitous act of cunning, spinning a tale on a golden spindle that brings a peace to the universe that humanity never truly knew on Earth. Voyager, by whistling in a "dark forest," may be the undoing of a violent ontology.

So the very initiative itself, for all its earthly inadequacies, is enough for peace — even for skeptics. It is not that the human has recentered itself as the agent of all meaning. The face of alterity — the alien, the other — is an anonymous call prior to our initiative, as Emmanuel Lévinas might say: it moves us first.[17] This is why the Voyager mission is as much an imaginary act of listening as it is one of communicating. Our musical response to the alien is an ethics of hospitality, as if we have already heard their call; it may be foolishly naive, but it is necessarily foundational if the indifferent landscape of a flat ontology is to resonate with a coherent rhythm.

Music means far more than we currently dare to imagine. If, as Raimon Panikkar states, "rhythm is another name for Being," then the Golden Record bears witness, however tentatively, to the very cosmic order in which it wanders.[18] As long as Voyager continues to journey into the unknown, it holds out the possibility that peace exists; "the launching of this bottle into the cosmic ocean says something very hopeful about life on this planet."[19] This gift is an initial, preliminary exploration — a probe — in search of an exchange that is far greater than anything the meager machine offers. And it might finally answer the reason for music.

★ ★ ★

Exchange is also the purpose behind this book. As with Voyager, we are in search of an exchange greater than anything these meager pages can offer in the hope that music might be an inter-relational envoy. In providing media theory and music theory as the two poles of our magnetic antennae, we hope to cause some creative clusters to come into friction and generate enough attraction and repulsion to create new waves of research with unending frequency modulations. We've come in peace, so now tune in, mind the gap, and make some noise. It's over to you.

~~THE END~~
REPEAT

Appendix

If the transparent film that functions
as a filter that rocks back and forth across
the artworks by Takahiro Kurashima
has been separated from this book, the
following grid can be copied onto a
transparent film and used as a substitute.

Technicians attaching the Golden Record to the Voyager space probe, 1977 (image: NASA/JPL).

Notes

INTRODUCTION

1. "Music theory" is often referred to in its shortened form as "theory" in this book. The word "theory" has different meanings in different disciplines. When it appears alone in these pages, it refers only to music theory.

2. Clearly, we are floating into posthuman territory here. There are many forms of posthumanism. In these pages, we are less concerned with the AI transhuman-cyborg variety than the decentering-anthropomorphic-dominance type. Our exploration focuses on what happens to music when the human is no longer in control of the materials of music. Music assumes a more thinglike, material, technologically mediated existence, with a vitality akin to what Jane Bennett calls "vibrant matter" (although, in this case, music is more a "vibrational matter") available to species beyond the human. See Jane Bennett, *Vibrant Matter: A Political Ecology of Things* (Durham, NC: Duke University Press, 2010).

3. Ian Bogost, *Alien Phenomenology, or What It's Like to Be a Thing* (Minneapolis: University of Minnesota Press, 2012), p. 32. Bogost is not referring to extraterrestrials in his title, although the theory fits, and we hope our creative spin honors the spirit of the book by extending it to the final frontier.

4. Graham Harman, *Guerrilla Metaphysics: Phenomenology and the Carpentry of Things* (Chicago: Open Court, 2005), p. 183. Harman's use of the term "noise" is more metaphorical than real, whereas, given the musical focus of this book, it is the other way around for us. For a more physical notion of background hum as an animating presence, see Lawrence Kramer, *The Hum of the World: A Philosophy of Listening* (Oakland: University of California Press, 2018).

5. Bennett, *Vibrant Matter.*

6. Niklas Luhmann, "The Cognitive Program of Constructivism and the Reality That Remains Unknown," in William Rasch, ed., *Theories of Distinction: Redescribing the Descriptions of Modernity* (Stanford, CA: Stanford University Press, 2002), p. 145.

7. See Levi R. Bryant, *Onto-Cartography: An Ontology of Machines and Media* (Edinburgh: Edinburgh University Press, 2014), pp. 54–62.

8. See Martin Heidegger, *Being and Time*, trans. John Macquarrie and Edward Robinson (New York: Harper & Row, 1962), H69–H364, and Graham Harman, *Tool-Being: Heidegger and the Metaphysics of Objects* (Chicago: Open Court, 2002).

9. The best-known and most influential retelling of this myth is in Boethius, *De institutione musica*, 1.10. See Anicius Manlius Severinus Boethius, *Fundamentals of Music*, ed. Claude V. Palisca, trans. and eds. Calvin M. Bower and Claude V. Palisca (New Haven, CT: Yale University Press, 1989), p. 17. The earliest surviving record of the myth is in Nicomachus of Gerasa, *Enchiridion harmonices*. See *The Manual of Harmonics of Nicomachus the Pythagorean*, ed. and trans. Flora R. Levin (Grand Rapids, MI: Phanes Press, 1994), p. 83.

10. Andreas Werckmeister, one of the last proponents of Pythagorean theory—albeit under the rule of Christendom, with God as the operator of a harmonious universe—describes humanity as "a tool" or "instrument." See Andreas Werckmeister, *Musicalische Paradoxal-Discourse: A Well-Tempered Universe*, trans. Dietrich Bartel (1707; Lanham, MD: Lexington Books, 2018), p. 26.

11. With the absorption of Pythagorean music theory into Christian theology, an element of temporality (since there is a genesis) enters the immutable numbers of celestial harmony and, in some cases, Creation explodes into being quite literally from a big bang: Hildegard of Bingen, in her *Scivias* of 1151 or 1152 (section 2.1.6), imagines the creation of the universe as an act of the Trinity, banging the materials into shape as if on an anvil and striking sparks into the atmosphere with blow after blow until heaven and earth are perfected by the hand of the divine workman. Pythagoras's forge becomes a divine workshop. See Hildegard of Bingen, *Scivias*, trans. Columba Hart and Jane Bishop (New York: Paulist Press, 1990), p. 152. Augustine, in *De musica*, also introduces temporality into the Pythagorean universe by turning its harmonic proportions into rhythmic relations. Augustine's musical universe is closer to the rhythm of dots that opens this introduction than to the fixity of Pythagorean numbers and so forms a bridge between the nonhuman music of Pythagoras's timeless cosmos and the space-time universe of our posthuman intergalactic music theory.

12. See Daniel K.L. Chua, *Absolute Music and the Construction of Meaning* (Cambridge: Cambridge University Press, 1999), pp. 29–113.

13. Werckmeister, *Musicalische Paradoxal-Discourse*, p. 65. The well-measured description fortuitously overlaps with the Wisdom of Solomon (11.20b, Apocrypha, NRSV), validating Werckmeister's theistic-Pythagorean universe.

14. There are, of course, exceptions, particularly the work of the music theorist Christopher F. Hasty in *Meter as Rhythm* (Oxford: Oxford University Press, 1997).

15. Object-oriented ontology (often abbreviated to OOO and aligned with speculative realism) is a current philosophy movement, loudly proclaimed by its leading proponent, Graham Harman, as "a theory of everything." Given our interest in a music theory of everything, OOO is a fellow space traveler on our intellectual journey. But note our use of the plural here—"ontologies." We are less concerned with the specific philosophical axioms that define this movement than the general orientation toward nonhuman objects in recent scholarship, not only in posthuman philosophies, but in many other disciplines, including media theory (e.g., Friedrich Kittler) and sociology of science (e.g., Bruno Latour). "Object-oriented" is a useful term because it describes succinctly what it means—an orientation toward objects. Our approach is, therefore, not a claim to being objective or to having direct access to the real, but an attitude or position that attempts to give objects their due. So when the term "object-oriented" appears in our text, it should be taken generally to mean ways of thinking that are oriented toward objects (which, of course, includes OOO). When OOO specifically is referred to, we will use the singular "ontology."

16. We have suppressed our philosophical urges into the occasional convoluted endnote. They are for the philosophically inclined.

17. A cosmic harmony is fundamentally an ontology of peace, with prelapsarian undertones that resonate, for example, in the theology of Hildegard of Bingen. In her letter to the prelates of Mainz of 1178 or 1179, she states that singing reconnects humanity to an Eden before the Fall. The team around Carl Sagan that created the Golden Record (notably Francis Drake, Ann Druyan, Tim Ferris, Jon Lomberg, and Linda Salzman Sagan) was well aware of the violent and destructive possibilities of alien civilizations; however, by sending music on a peace mission, Voyager amplifies the distant echoes of an ancient cosmology. Or to modernize the conditions of cosmic harmony, we can imagine Voyager as someone whistling in the dark (or in the words of Cixin Liu's influential science-fiction trilogy, *Three-Body Problem*, whistling in a "dark forest"). Whistling in the dark gestures to a faint, but necessary peace in a scary universe. Deleuze and Guattari's musical ontology opens with an analogous sound: "A child, in the dark, gripped with fear, comforts himself by

singing under his breath." This barely audible song is the "beginnings of order in chaos"; with its constant repetition, this singing under the breath illustrates the full-blown concept of the *ritournelle* (a repeated ditty translated as "refrain") that bursts forth as if from a "black hole" and "fan[s] out to the sphere of the cosmos." Voyager, in this sense, hums our song in the dark, as if bringing a slice of our territory — our home planet — into unknown territories, both as consolation and as peace envoy. See Gilles Deleuze and Félix Guattari, *A Thousand Plateaus: Capitalism and Schizophrenia*, trans. Brian Massumi (Minneapolis: University of Minnesota Press, 1987), pp. 311–12.

CHAPTER ONE: MANIFESTO

1. Speaking in very general terms, much of the exciting work in current music theory happens at the margins, while the center, which has lost its former focus, soldiers on. Will the periphery flip to the center?

2. Richard Taruskin, "A Myth of the Twentieth Century: *The Rite of Spring*, the Tradition of the New, and 'the Music Itself,'" *Modernism/Modernity* 2.1 (2002), pp. 1–26. Although we will recover the notion of absolute music as a music that "absolves" (literally "loosens") itself from the Earth as an object in space, its earthly counterpart since about 1800 has tended to "absolve" itself from the rest of the sciences and humanities, if not from musicology, and has been the ideological justification for music theory's isolationism. In fact, "the absolute" in general sides with an object-focused view, as Jane Bennett points out; so if recalibrated, an absolute music can be the cure for our diagnosis of music theory's "absolute" problem. The poison is the cure! See Jane Bennett, *Vibrant Matter: A Political Ecology of Things* (Durham, NC: Duke University Press, 2010), pp. 2–3. On the ideology of absolute music, see Carl Dahlhaus, *The Idea of Absolute Music*, trans. Roger Lustig (Chicago: University of Chicago Press, 1989); Mark Evan Bonds, *Absolute Music: The History of an Idea* (New York: Oxford University Press, 2014); and Daniel K. L. Chua, *Absolute Music and the Construction of Meaning* (Cambridge: CambridgeUniversity Press, 1999).

3. These numbers, adorned with a triangular hat, denote the scale degrees of the *Urlinie*, which can descend from $\hat{3}$ to $\hat{1}$ or $\hat{5}$ to $\hat{1}$ or $\hat{8}$ to $\hat{1}$, in the music theory of Heinrich Schenker.

4. We are not, of course, the first to critique the field of music theory. For trenchant critiques see — among many others — Kevin Korsyn, *Decentering Music: A Critique of Contemporary Musical Research* (New York: Oxford University Press, 2003); Andrew dell'Antonio, ed., *Beyond Structural Listening: Postmodern Modes of Hearing* (Berkeley:

University of California Press, 2004); Joshua Mailman, "Cybernetic Phenomenology of Music, Embodied Speculative Realism, and Aesthetic-Driven Techné for Spontaneous Audio-Visual Expression," *Perspectives of New Music* 54.1 (2016), pp. 5–95; or Kofi Agawu, "Tonality as a Colonizing Force in Africa," in Ronald Radano, ed., *Audible Empire: Music, Global Politics, Critique* (Durham, NC: Duke University Press, 2018), pp. 334–55.

5. Among 115 images encoded on the golden disc is a photograph entitled "Demonstration of Licking, Eating, and Drinking," showing members of NASA's research team biting into a toasted tuna sandwich, licking an ice-cream cone, and pouring a jug of water down a throat; if decoded by an alien, the image may convey the wrong idea of how humans consume music.

6. In an episode of *Saturday Night Live* (season 3, episode 64, originally aired in 1978), Steve Martin's character, a psychic named Cocuwa, announces that aliens had responded to the Golden Record: "Send more Chuck Berry!"

CHAPTER TWO: BLUEPRINT

The epigraph from Friedrich Hölderlin is attributed to him in Bettina von Arnim and Karoline von Arnim, *Die Günderode* (Frankfurt am Main: Insel, 1982), p. 331.

1. See Elizabeth H. Margulis, *On Repeat: How Music Plays the Mind* (New York: Oxford University Press, 2014), pp. 15–18. Language can be perceived as music if a spoken phrase is repeated sufficiently, as in Diana Deutsch's speech-to-song illusion: Diana Deutsch, Trevor Henthorn, and Rachael Lapidis, "Illusory Transformation from Speech to Song," *Journal of the Acoustical Society of America*, 129.4 (2011), pp. 2245–52. Even noise or a sound-fragment association with a source or referent, as Pierre Schaeffer points out, loses its "narrative" component and becomes music when repeated: "Repeat the same sound-fragment twice: There is no longer event, but music." See Pierre Schaeffer, *In Search of a Concrete Music*, trans. Christine North and John Dack (Berkeley: University of California Press, 2012), p. 13.

2. Bernhard Siegert dedicates a major section of his magisterial *Passage des Digitalen: Zeichenpraktiken der neuzeitlichen Wissenschaften 1500–1900* (Berlin: Brinkmann und Bose, 2003) to the mathematician Joseph Fourier, who showed that all wave forms, no matter how complex, can be broken down into simple sinusoids. In Fourier's universe, everything is vibration.

3. See Catherine Pickstock, *Repetition and Identity* (Oxford: Oxford University Press, 2013), pp. 43–66.

4. Søren Kierkegaard, *Repetition and Philosophical Crumbs*, trans. M. G. Piety (Oxford: Oxford University Press, 2009), pp. 4. See note 12 below.

5. Johann Gottfried Herder, *Werke*, vol. 8, *Schriften zu Literatur und Philosophie, 1792–1800*, ed. Hans Dietrich Irmscher (Frankfurt am Main: Deutscher Klassiker Verlag, 1998), pp. 705 and 812.

6. The claims here are indebted to object-oriented ontology; see Graham Harman, *Object-Oriented Ontology: A New Theory of Everything* (London: Penguin, 2018), pp. 61–102.

7. See Henry Stobart, "Devils, Dreams, and Desire: Siren Traditions and Musical Creation in the Central Southern Andes," in Linda Phyllis Austern and Inna Naroditskaya, eds., *Music of the Sirens* (Bloomington: Indiana University Press, 2006), pp. 119–21.

8. Alfred Gell, *Art and Agency: An Anthropological Theory* (Oxford: Oxford University Press, 1998), p. 80.

9. Music-theoretical paradigms tend to posit higher-order unities—formal models, hierarchical tree structures, teleological progressions, long-term tension-resolution dialectics.

10. See Daniel Heller-Roazen, *The Fifth Hammer: Pythagoras and the Disharmony of the World* (New York: Zone Books, 2011), pp. 61–88. In his discussion of metrical feet in *Elementa Rhythmica*, Aristoxenus of Tarentum points to this concept in paragraph 18: "It is apparent that there cannot be a foot of one time interval, since indeed one signal does not make a distribution of time." The translation is from Christopher C. Marchetti, "Aristoxenus' *Elements of Rhythm*: Text, Translation, and Commentary with a Translation and Commentary on *POxy 2687*," PhD diss., Rutgers University, 2009, p. 70.

11. For a similar argument, see Paul North, *Bizarre-Privileged Items in the Universe: The Logic of Likeness* (New York: Zone Books, 2021).

12. As Søren Kierkegaard underlines, repetition is not "the same": repetition is "the new." He states in *Repetition*, p. 19: "The dialectic of repetition is easy, because that which is repeated has been, otherwise it could not be repeated; but precisely this, that it has been, makes repetition something new." Kierkegaard is speaking of life events, but since everything repeats, music is a way of articulating and reflecting, in its own terms, what the philosopher calls "the earnestness of existence" that an immanent, future-orientated, and infinitely variable repetition expresses; see Kierkegaard, *Repetition*, p. 4. In this sense, Kierkegaard is foundational for Deleuze, for whom repetition also swerves away from sameness and identity toward radical difference. See Gilles Deleuze, *Difference and Repetition*, trans. Paul Patton (New York: Columbia University Press, 1994). To complete our brief genealogy of repetition, both Kierkegaard and Deleuze are indebted to Leibniz; see note 20 below.

13. See Pickstock, *Repetition and Identity*, p. 23.

14. The terms "archi-textural" is coined by Henri Lefebvre, *The Production of Space*, trans. Donald Nicholson-Smith (Oxford: Blackwell, 1991), pp. 117–18. Also see Tim Ingold, for whom social reality is a meshwork of woven lines and knots: *Lines: A Brief History* (London: Routledge, 2016), and *The Life of Lines* (London: Routledge, 2015). As Ingold points out, many cultures understand the habitation of time and space as a weave. Despite the posthuman slant of IMTE, there is an anthropology knotted in its texture.

15. Musical forms of all kinds, as James Hepokoski's work underlines, are rotations. See, for example, James Hepokoski, *Sibelius: Symphony No. 5* (Cambridge: Cambridge University Press, 1993), pp. 23–26 and 58–84, and Hepokoski, "Structure, Implication, and the End of *Suor Angelica*," *Studi pucciniani* 3 (2004), pp. 241–64.

16. These different frequencies can be heard on the Golden Record: a groove in "Johnny B. Goode," a drone in Bhairavi: "Jaat Kahan Ho," a ritornello in Bach's Brandenburg Concerto no. 2, movement 1. Positing music's ground as a glissandolike spectrum of frequencies gives a whole new spin to the concept of transposition. Jon Lomberg, the design director of the Golden Record's team, included a Bach fugue because the logic of its internal structure would remain intact at whatever frequency it is played. It could be "played in long or short wavelengths and the frequency intervals would remain in the same ratio. So a fugue is a fugue, whether in UV, sound, or gamma ray wavelengths." It is a matter of scale. To understand Earth's music, alien listening would require both a refolding and a "rescaling" of the spectrum to accommodate the aliens' musical receptors. The quote is from a personal correspondence dated August 8, 2020.

17. In this sense, IMTE is a "monadology," and we use the term "point of view" technically as a Leibnizian concept: each one of us, like a monad, registers the totality of repetition as a blur and can focus clearly on a specific point only from the perspective of our disciplinary understanding. By coordinating our vision in a modular fashion — somewhat like a compound eye — we might approximate toward a rhythmic theory of everything. Your point of view is precisely what you signed up for on the commitment form.

18. See Gottfried Wilhelm Leibniz, *Die philosophischen Schriften*, vol. 7, ed. C. I. Gerhardt (1875–90; Hildesheim: Olms, 1960–61), p. 503.

19. In book 19 of the *Odyssey*, when Penelope recounts the narrative of the weave to Odysseus, the hero who has just returned in disguise and is meeting with Penelope for the first time: he is a stranger.

20. Deleuze, *Difference and Repetition*, p. 293. The influence of Deleuze is clearly evident

in our theory of music; in particular, his book *The Fold: Leibniz and the Baroque* underlines many of our premises, mainly because our work, with its infinite gradations of repetition, has an acute Leibnizian slant in its approach to music and the cosmos. There are various parallels between IMTE and the philosophy of Leibniz: IMTE is a music theory of *sufficient reason* (encompassing all possible rhythms) from a *point a view* (expressed as pieces of time/infinity) in accordance with the *principle of indiscernibles* (in which no two pieces are the same) within a universe of *preestablished harmony* where communication is possible between monads without prior knowledge (namely, rhythm as the resonance for alien contact). Deleuze, in *Difference and Repetition* (1968), builds on these principles and revisits them twenty years later in *The Fold* (1988). We share with Deleuze a process-based approach, a fascination with the generative energy of pleats and folds, and an eclectic, all-inclusive vision that expands music beyond the human, and we use some of the same terminology—"fold," "difference," "repetition," "rhythm," "event." However, they are deployed in divergent ways determined more by *how music goes* than by a univocal philosophy of morphogenesis that issues forth from a metaphysics of difference. We come in peace—Deleuze comes in chaos! In fact, on major points, we often differ, as is evident in our emphasis on music as weaving; for Deleuze, weaving represents a striated space-time in music, a form of hierarchical governance and law-bound terrain at odds with his promotion of smooth, nonformal, and amorphous assemblages. Our conception of weaving takes the notion of repetition and difference in another direction and is not conditioned by Deleuze's striated definition. Ultimately, Deleuze's fabric of thought is different from ours, and any attempt to fit our terms into his definitions would be confusing, which is why such a central influence remains at the margins of our text. On weaving, see Gilles Deleuze and Félix Guattari, *A Thousand Plateaus: Capitalism and Schizophrenia*, trans. Brian Massumi (Minneapolis: University of Minnesota Press, 1987), pp. 474–78, and Gilles Deleuze, *The Fold: Leibniz and the Baroque*, trans. Tom Conley (Minneapolis: University of Minnesota Press, 1993).

21. See Bill Brown, "Thing Theory," *Critical Inquiry* 28.1 (2001), pp. 1–22.

22. Gary Tomlinson, *One Million Years of Music: The Emergence of Human Modernity* (New York: Zone Books, 2015), p. 74.

23. See Theodor W. Adorno and Max Horkheimer, *Dialectic of Enlightenment*, trans. John Cumming (London: Verso, 1979).

24. See Denis Diderot, *D'Alembert's Dream* (1769), in *Diderot's Selected Writings*, ed. Lester G. Crocker, trans. Derek Coltman (New York: Macmillan, 1966), pp. 187–88; *nervus*

means literally string. Figure 2.5, from Hermann von Helmholtz, *On the Sensations of Tone* (New York: Dover, 1954), p. 139, was first published in 1863 as *Die Lehre von den Tonempfindungen als physiologische Grundlage für die Theorie der Musik*. On the cochlea as piano, see Hermann von Helmholtz's lecture "On the Physiological Causes of Musical Harmony (1857)," in David Cahan, ed., *Science and Culture: Popular and Philosophical Essays* (Chicago: University of Chicago Press, 1995), p. 60.

CHAPTER THREE: SENDER

1. On media archaeology, see Jussi Parikka, *What is Media Archaeology?* (Cambridge: Polity, 2012); Jussi Parikka and Erkki Huhtamo, eds., *Media Archaeology* (Berkeley: University of California Press, 2011), and Wolfgang Ernst, *Digital Media and the Archive*, trans. Jussi Parikka (Minneapolis: University of Minnesota Press, 2012).

2. The making of the Golden Record is documented in Carl Sagan, Francis D. Drake, Ann Druyan, Timothy Ferris, Jon Lomberg, and Linda Salzman Sagan, *Murmurs of Earth: The Voyager Interstellar Record* (New York: Random House, 1978).

3. The folk song collector Alan Lomax urged the team to go beyond the European musical tradition in their selection of music. See *Murmurs of Earth*, p. 16. The concept of World Music has been treated extensively by ethnomusicologists, perhaps most comprehensively in Philip Bohlman, ed., *The Cambridge History of World Music* (Cambridge: Cambridge University Press, 2014). Among the reflections on the concept by and for nonspecialists, those by Peter Sloterdijk, *Kopernikanische Mobilmachung und Ptolemäische Abrüstung: Ästhetischer Versuch* (Frankfurt am Main: Suhrkamp, 1987), pp. 77–120; Timothy Brennan, "World Music Does Not Exist," *Discourse* 23.1 (2001), pp. 44–62; and Katie Trumpener, "World Music World Literature: A Geopolitical View," in Haun Saussy, ed., *Comparative Literature in an Age of Globalization* (Baltimore: Johns Hopkins University Press, 2006), pp. 183–202, are noteworthy.

4. Carl Sagan, *The Cosmic Connection: An Extraterrestrial Perspective* (Garden City, NY: Anchor Books, 1980), p. 167.

5. In a moment of truly poetic timing, the famous "WOW" signal, a strong radio signal that might have stemmed from an artificial extraterrestrial source, was recorded only five days before the launch of Voyager 2, on August 15, 1977.

6. See Stephen Webb, *If the Universe Is Teeming with Aliens, Where Is Everybody?* (Heidelberg: Springer, 2015).

7. Sagan et al., *Murmurs of Earth*, p. 11.

8. The publication story of this book is complicated. Written by Sagan's Soviet counterpart, Iosif Shklovskii, and first published in 1962 under the title Вселенная, жизнь, разум (Universe, life, intelligence), the 1966 English translation of Shklovskii's book was edited and expanded by Carl Sagan, who assumed the title of coauthor.

9. Sagan et al., *Murmurs of Earth*, p. 163, and more explicitly in Timothy Ferris, "How the Golden Record Was Made," *New Yorker*, August 20, 2017.

10. Anthony Michael Morena, *The Voyager Record: A Transmission* (Brookline, MA: Rose Metal Press, 2016), p. 135.

11. Roland Barthes, *Camera Lucida: Reflections on Photography*, trans. Richard Howard (New York: Hill & Wang, 2010).

12. Evander Price, "Future Monumentality," PhD diss., Harvard University, 2019.

13. Jimmy Carter's statement onboard Voyager.

14. Ann Druyan in Scott, *The Vinyl Frontier: The Story of the Voyager Golden Record* (London: Bloomsbury, 2019), p. 200.

15. Timothy Ferris, in Sagan et al., *Murmurs of Earth*, p. 203.

16. Jimmy Carter's statement on Voyager: "We human beings are still divided into nation states, but these states are rapidly becoming a single global civilization."

17. On time, lateness, and Beethoven's late style, see Theodor W. Adorno's classic essay of 1934, "Spätstil Beethovens," in *Beethoven: The Philosophy of Music*, ed. Rolf Tiedemann, trans. Edmund Jephcott (Cambridge: Polity, 1998), pp. 123–26; also see Daniel K.L. Chua, "In Time with the Late Style," preface in Leo Ou-fan Lee and Henry Shiu, 【諸神的黃昏】– 晚期風格的跨學科對談: *In Time with the Late Style* (Hong Kong: Oxford University Press, 2019). On the Cavatina, see Daniel K.L. Chua, "Beethoven's Other Humanism," *Journal of the American Musicological Society* 62.3 (2009), pp. 571–645. For a perceptive account of nostalgia, see Svetlana Boym, *The Future of Nostalgia* (New York: Basic Books, 2004).

18. Friedrich Nietzsche, *The Gay Science*, trans. Walter Kaufmann (New York: Vintage, 1974), p. 262.

19. Ann Druyan, "Dear Voyagers: How Your Billion-Year Journey Carries True Love," *National Geographic*, July 10, 2019. Druyan's "love letter" vision of the Voyager mission is personal: she and Carl Sagan fell in love during the project. The Golden Record is even sealed by Druyan with a kiss—or, rather, the sound of a kiss in the sound essay—although, intriguingly, it was planted on her cheek by the producer of the record, Timothy Ferris, who was her boyfriend when the project started. As with the love interest in many a tale, hers has taken on mythic proportions, often eclipsing the curatorial team that lovingly

assembled each piece on the record. Love, in many ways, does underlie the Voyager mission, but it is not so much Druyan's romance, but the creative and conceptual work of the entire team that is the labor of love in this letter to the universe.

20. Mixtapes and playlists have long been the stuff of the romantic-comedic genre, from Nick Hornby's best-selling novel *High Fidelity* (1995) to Rob Sheffield's *Love Is a Mix Tape* (2007). But there's more to the unassuming tape than meets the eye—and not only the ubiquitous artwork in the cover. Academics have only begun to scratch the surface of this cultural phenomenon. See Bas Jensen, "Tape Cassettes and Former Selves: How Mix Tapes Mediate Memories," in Karin Bijsterveld and José van Dijck, eds., *Sound Souvenirs: Audio Technologies, Memory, and Cultural Practices* (Amsterdam: Amsterdam University Press, 2009), pp. 43–54.

21. Steven Pinker, *How the Mind Works* (New York: Norton, 1997), p. 534.

22. Ian Cross, "Is Music the Most Important Thing We Ever Did?: Music, Development, and Evolution," in Suk Won Yi, ed., *Music, Mind, and Science* (Seoul: Seoul National University Press, 1999), pp. 10–39.

23. Sebastian von Hoerner, "Universal Music?," *Psychology of Music* 2.2 (1974), pp. 18–28.

24. Sebastian von Hoerner, *Sind wir allein? SETI und das Leben im All* (Munich: C H. Beck, 2003), p. 143. See also Sagan et al., *Murmurs of Earth*, pp. 13–14.

25. Simon was a late 1970s electronic game in which players were required to remember and repeat sequences of tones and lights. Success led to series that became progressively longer and more complex.

26. This list is corrected, following Jonathan Scott, *The Vinyl Frontier: The Story of the Voyager Golden Record* (London: Bloomsbury, 2019), pp. 271–72. The list can be heard on YouTube, SoundCloud, and Spotify.

27. The exclusion of the Beatles has been debated quite a lot among aficionados. The most complete discussion is found in Scott, *Vinyl Frontier*, pp. 144–46. Jon Lomberg, the design director, in an email conversation (August 11, 2020), confirms that EMI turned down the request to include the Beatles, and explains that Bob Dylan, the Rolling Stones, and Elvis were discussed as backups, but Chuck Berry won out.

28. See Sagan et al., *Murmurs of Earth*, pp. 23–25.

29. Susan Stewart, *On Longing: Narratives of the Miniature, the Gigantic, the Souvenir, and the Collection* (Durham, NC: Duke University Press, 1992), p. 153.

30. Sagan et al., *Murmurs of Earth*, p. 21.

31. See John Thompson, "06: Flowing Streams, Appendix: Chart Tracing Liu Shui," http://www.silkqin.com/02qnpu/07sqmp/sq06ls.htm#chart.

32. The team steered clear of vocal music when compiling the samples of Western classical music: the absence of Italian opera has long been noted, and the only piece of vocal music is the aria of the Queen of the Night, whose extreme vocal demands overshadow language and text. (While the team was always concerned that the words of texted music might cause offense, they clearly did not look very carefully at this aria, whose sentiment is outright inhuman[e].)

33. Alice Gorman, "Beyond the Morning Star: The Real Tale of the Voyager's Aboriginal Music," *The Conversation*, October 2, 2013, http://theconversation.com/beyond-the-morning-star-the-real-tale-of-the-voyagers-aboriginal-music-18288.

34. Scott, *Vinyl Frontier*, pp. 91–93, identifies a number of these misattributions.

35. Sagan et al., *Murmurs of Earth*, p. 188.

36. Derrida's famous dictum *il n'y a pas de hors-texte*—"there is nothing outside the text"—suggests itself here. Jacques Derrida, *Of Grammatology*, corrected edition, trans. Gayatri Chakravorty Spivak (Baltimore: Johns Hopkins University Press, 1998), p. 158. The textuality of the Golden Record will be the topic of Chapter 5.

37. Sagan et al., *Murmurs of Earth*, 12. The suggestion was made by Barnie Oliver. From a purely technical perspective, magnetic tape brings with it some material problems and, exposed to radiation and extreme temperatures in outer space, would not be very durable.

38. Ibid., p. 13. This suggestion came from Lewis Thomas.

39. Ibid., p. 163. The selection of the music remained torn between the two poles throughout the project. According to Jon Lomberg in a personal communication, he and Drake thought more strictly in terms interstellar communication following universal principles; they chose music on the basis of what may be most accessible to an alien audience. But for Sagan and the others on the team, it seemed more important to pick music that had the strongest possible emotional impact on *humans* and trust to chance whether aliens would like humanity's greatest hits.

40. Spiegel programmed a computer with updated scientific data on the motion of the six planets known to Kepler.

CHAPTER FOUR: TRANSMISSION

1. Niklas Luhmann, "The Improbability of Communication," *International Social Science Journal* 23.1 (1981), pp. 121–32.

2. Carl Sagan, Francis D. Drake, Ann Druyan, Timothy Ferris, Jon Lomberg, and

Linda Salzman Sagan, *Murmurs of Earth: The Voyager Interstellar Record* (New York: Random House, 1978), p. 7.

3. Ibid., p. 11.

4. Ibid., p. 6.

5. Silicon has qualities similar to carbon and might be able to fulfill the structural role of carbon in other ecosystems.

6. Sybille Krämer, "The Cultural Techniques of Time Axis Manipulation: On Friedrich Kittler's Conception of Media," *Theory, Culture & Society* 23.7–8 (2006), pp. 93–109.

7. This is not a hypothetical issue. The US news program *60 Minutes* attempted to produce a feature on the Golden Record in which they would play music from the copy of the Golden Record that's still on this planet—at NASA's Jet Propulsion Lab in Pasadena, California. But they found that the hole in the center of the disc was not a standard size and could not be played on a standard turntable. This unfortunate situation was soon reduced to an alarming rumor: "The Golden Record cannot be played!" While technically correct, that is somewhat misleading.

8. To put numbers to it, the frequency of the emitted radiation is 1,420,405,751.7667 ±0.0009 Hz, which makes the wave almost precisely 21 centimeters long. This is why the hydrogen line is also known as the "twenty-one-centimeter line." NASA has used the line as a universal unit of length, famously in the context of the Pioneer plaque, but it functions equally well as a universal unit of time.

9. Sagan et al., *Murmurs of Earth*, p. 12.

10. See Jussi Parikka, *What is Media Archaeology?* (Cambridge: Polity, 2012).

11. To be sure, this diagram is schematic. The actual production process, for instance, involves multiple stages of recording, engraving, and copying the groove that are simplified here into one. And the listening stage always involves the possibility of not perceiving.

12. Emily Dolan, "Toward a Musicology of Interfaces," *Keyboard Perspectives* 5 (2012), pp. 1–13.

13. Friedrich Kittler, *Gramophone, Film, Typewriter*, trans. Geoffrey Winthrop-Young and Michael Wutz (Stanford, CA: Stanford University Press, 1999). Kittler's perspective, which tells a specifically Western story of writing, has come under some criticism. In its universalizing tendency, the nineteenth-century gramophone has also been recognized as a crucial device in the service of colonial expansion. See, for instance, Ana Maria Ochoa, "Exchange: The Geopolitics of Ethnographic Recordings, Music, and Sound," in Naomi

Waltham-Smith and Alexander Rehding, eds., *A Cultural History of Western Music*, vol. 5: *The Industrial Age* (London: Bloomsbury, forthcoming).

14. Despite its many problems, the classic essay by R. H. Van Gulik, "The Lore of the Chinese Lute: An Essay in Ch'in Ideology," *Monumenta Nipponica* 1.2 (1938), pp. 386–438, remains the most comprehensive introduction in English. On the problems of converting notation, see also Alexander Rehding, "The Dream of a Universal Notation: An Appreciation," *Analytical Approaches to World Music* 8.2 (2020), pp. 314–23.

15. The concept of ekphrastic hope appears in W. T. J. Mitchell's "Ekphrasis and the Other," in *Picture Theory: Essays on Verbal and Visual Representation* (Chicago: University of Chicago Press, 1994), pp. 151–81. When Walter Kaufmann muses about the "vagueness" of "Oriental" notational systems, he puts exactly this problem on display: every act of transcription requires a certain ekphrastic hope. See his *Musical Notations of the Orient* (Indianapolis: Indiana University Press, 1967). There is no *hors-notation*, no Archimedean pivot point outside of forms of representation. See also note 17 below.

16. Not an unmediated impression of the sound wave, to be sure: what comes between the sound wave and the record groove is a whole apparatus of record production. See Jonathan Sterne, *The Audible Past: Cultural Origins of Sound Reproduction* (Durham, NC: Duke University Press, 2003).

17. As Derrida almost put it, *il n'y a pas de hors-enrégistration*—there is nothing outside of the recording. His famous line about the specific logic of text—or rather "arche-writing," as Derrida would call it—becomes, in Haun Saussy's helpful paraphrase, an admonition that "there is no extratextual, tipped-in illustration (*planche hors-texte*) that I can send you to, no stone I can tell you to kick, that would not obey a textual logic or testify to the textual condition." Haun Saussy, "Exquisite Cadavers Stitched from Fresh Nightmares: Of Memes, Hives, and Selfish Genes," in Haun Saussy, ed., *Comparative Literature in an Age of Globalization*, p. 33. Arche-writing is always prior to and around any specific instance of the text, and this axiom extends, in Kittler's adaptation of Derrida, to the technological mediation of music.

18. NASA felt, after the UN secretary general had been invited to include a spoken address on the Golden Record, that it could not but invite the US president, Jimmy Carter, as well, to include a greeting. NASA, after all, was funded by US taxpayers. President Carter's office chose to submit an address in typescript, rather than a voice recording, for the Golden Record. Apparently he did not want to project his pronounced Southern accent into the universe. See Jonathan Scott, *The Vinyl Frontier: The Story of the Voyager Golden Record* (London: Bloomsbury, 2019), p. 190.

19. We pick up a speculation here by Michael Anthony Morena, *The Voyager Record: A Transmission* (Brookline, MA: Rose Metal Press, 2016), p. 45.

20. Ron Barry, "How to Decode the Images on the Voyager Golden Record," *Boing Boing*, September 5, 2017, https://boingboing.net/2017/09/05/how-to-decode-the-images-on-th.html. The blog post was perfectly timed to serve as a birthday celebration for Voyager 1.

21. On the "digital" and the "analog" see Ben Peters, ed., *Digital Keywords: A Vocabulary of Information Society and Culture* (Princeton, NJ: Princeton University Press, 2016). To be precise, at eight seconds per image, the 115 images only take up about fifteen and three-quarters minutes playing time, about a quarter of the slowed-down record side. And to be even more precise, the side that faces inward, toward the main body of the record, is the side that holds the images, because this side is less likely to be damaged on its journey by micrometeorites. See Sagan et al., *Murmurs of Earth*, p. 234.

22. For a critique of Kittler's association of the gramophone with the Lacanian Real, see Mark B. N. Hansen, "Symbolizing Time: Kittler and Twenty-First Century Media," in Stephen Sale and Laura Salisbury, eds., *Kittler Now: Current Perspectives in Kittler Studies* (Cambridge: Polity, 2015), pp. 210–37.

23. Samuel Pepys, *Diary of Samuel Pepys: A New and Complete Transcription*, ed. Robert Latham and William Matthews (London: Bell, 1970), entry for August 8, 1666. And even earlier, Leonardo da Vinci and the circle around Galileo Galilei seem to have been aware of this phenomenon without, however, exploring it at a deeper level.

24. For more on this question, see Alexander Rehding, "Of Sirens Old and New," in Jason Stanyek and Sumanth Gopinath, eds., *The Oxford Handbook of Mobile Music Studies, Volume 2* (New York: Oxford University Press, 2014), pp. 77–106.

25. Karlheinz Stockhausen, "Four Criteria of Electronic Music," in *Stockhausen on Music: Lectures and Interviews*, ed. Robin Maconie (London: Marion Boyars, 1989), pp. 91–92.

26. For more on Opelt, see Alexander Rehding, "Opelt's Siren and the Technologies of Musical Hearing," in Viktoria Tkaczyk, Mara Mills, and Alexandra Hui, eds., *Testing Hearing: The Making of Modern Aurality* (New York: Oxford University Press, 2020), pp. 131–57.

27. Charles Cagniard de la Tour, "Sur la sirène, nouvelle machine d'acoustique pour mesurer les vibrations de l'air qui constituent le son," *Annales de la physique et chimie* 12 (1819), p. 171.

28. One of the first lessons that media archaeology teaches is that the search for a "first" is almost always fruitless and boring. This case is no exception. Opelt was preceded in this realization by Leonhard Euler and Johann Sulzer, who both speculated on this principle,

without, however, putting it into experimental practice. See Alexander Rehding, "Consonance and Dissonance," in Alexander Rehding and Steven Rings, eds., *The Oxford Handbook of Critical Concepts in Music Theory* (New York: Oxford University Press, 2019), pp. 441–47.

29. There is no (human) music that applies Opelt's theory in a systematic fashion. If we dig deep, we can come up with an extremely short and very diverse canon of compositions that gesture in this direction: Henry Cowell's youthful *Quartet Romantic* (1915) encodes a secret four-part Bach chorale in the rhythmic structure of the individual parts; the glissandos in Moby's "Thousand" and Stockhausen's *Kontakte* mentioned earlier. To a certain extent, Peter Ablinger's *Letter from Schoenberg* (2008) can also be added to this canon of recursivity, because it recreates the partials that make up the timbres of the speaking voice as complex harmonic textures performed on keys of the piano as a second-order timbre based on an even more complex harmonic structure.

30. Sybille Krämer is instrumental in highlighting the significance of time axis manipulation in her reading of Kittler; her important article "The Cultural Techniques of Time Axis Manipulation" elevates it to the central principle of Kittler's media theory. Some musical aspects of time axis manipulation are further explored in Alexander Rehding, *Beethoven's Symphony No. 9* (New York: Oxford University Press, 2017).

31. Birgit Schneider, *Textiles Prozessieren* (Zurich: Diaphanes, 2007) explores the media theory of the Jacquard loom.

32. Opelt's theory falters on the rigid spiral recursivity of his siren disc, which limits his theory in practical terms. His rotational logic cannot get beyond one single sound. The parallelisms between rhythm, pitch, phrase structure, and form are only theoretical positions. To turn these sophisticated positions into a sounding reality, Opelt would require an iterative program in the style of a Jacquard-style punch card, but that would require a different sounding mechanism, much closer to a modern tape player. Cowell's *Quartet Romantic* would be such a program. And — who knows? — there may be some ears out in the universe for whom the transformation from polyrhythms to four-part harmonies is not a complex task of transcription, but an automatic sensory transduction.

33. In a Necker cube, a drawing of a rhomboid first published in 1832 by Swiss crystallographer Louis Albert Necker, there are no clues to its orientation, so it can be interpreted to have either the lower-left or the upper-right square as its front side.

34. See especially Krämer, "Time Axis Manipulation."

35. Wolfgang Ernst, *Im Medium erklingt die Zeit: Technologische Tempor(e)alitäten* (Berlin:

Kulturverlag Kadmos, 2015), p. 218. Ernst's revised English translation of this book, *Sonic Time Machines, Explicit Sound, Sirenic Voices, and Implicit Sonicity* (Amsterdam: Amsterdam University Press, 2016), catapults this point into the title.

CHAPTER FIVE: RECEIVER

1. Drake has subsequently simplified the equation, arguing that most factors are either close to 1 or cancel each other out. The simple Drake equation is $N = L$. He even used this simplified Drake equation for his license plate: NEQUSL. Since he assumes the life cycle of a civilization L equals 10,000, this is also Drake's number of civilizations in the galaxy. Others disagree. Sebastian von Hoerner estimates between fifty thousand and one million advanced civilizations in our galaxy, and he assumes that the average distance is between 100 and 1,000 light years. Astrobiologist Sara Seager has recently developed a parallel equation, usually given as $N = N^{*} \cdot f_Q \cdot f_{HZ} \cdot f_o \cdot f_L \cdot f_S$, which bypasses alien radio technology and instead focuses the search on biosignatures, gases produced by living organisms.

While the visualization in Figure 5.1 features prominently on the internet, it is hard to identify a point of origin. The original source seems to be Colin A. Houghton of noeticscience.co.uk, though the original post has been taken down.

2. These and other possible aliens are imagined in Michael Anthony Morena, *The Voyager Record: A Transmission* (Brookline, MA: Rose Metal Press, 2016).

3. On the Fermi paradox, see Stephen Webb, *If the Universe Is Teeming with Aliens, Where Is Everybody?* (Heidelberg: Springer, 2015).

4. Stefan Helmreich, "Music for Cochlear Implants," in Emily Dolan and Alexander Rehding, eds., *The Oxford Handbook of Timbre* (New York: Oxford University Press, 2020).

5. For recent developments in cross-species approaches to music cognition, see especially Aniruddh D. Patel, "Evolutionary Music Cognition: Cross-Species Studies," in P. Jason Rentfrow and Daniel Levitin, eds., *Foundations in Music Psychology: Theory and Research* (Cambridge, MA: The MIT Press, 2019), pp. 459–501.

6. The classic text on bat audition and echolocation is Donald R. Griffin, *Listening in the Dark: The Acoustic Orientation of Bats and Men* (New York: Dover, 1974).

7. Thanks go to Bevil Conway and his lecture "Rats, Bats, and Platypus," on these creatures' perceptual worlds.

8. Jakob von Uexküll, *A Foray into the Worlds of Animals and Humans: With a Theory of Meaning*, trans. Joseph D. O'Neil (Minneapolis: University of Minnesota Press, 2010).

9. See Richard Grusin, ed., *The Nonhuman Turn* (Minneapolis: University of Minnesota Press, 2015).

10. Graham Harman underlines the importance of metaphors in his *Object-Oriented Ontology: A New Theory of Everything* (London: Penguin, 2018), pp. 61–102.

11. Vilém Flusser and Louis Bec, *Vampyroteuthis Infernalis: A Treatise, with a Report by the Institut Scientifique de Recherche Paranaturaliste*, trans. Valentine Pakis (Minneapolis: University of Minnesota Press, 2012).

12. See Peter Godfrey-Smith, *Other Minds: The Octopus, the Sea, and the Deep Origins of Consciousness* (London: Collins, 2016).

13. The numbers vary somewhat between experiments. Squid audition seems to be focused a little more broadly, between 100 and 1,500 Hz.

14. We needn't be too concerned about the upper end. Experiments suggest that this may be related to the dwelling place of the octopus at the bottom of the sea, where waves of 1000 Hz—which are 1.5 meters long—get blocked in the water by obstacles larger than one wavelength.

15. Sam V. Norman-Haignere, Nancy Kanwisher, Josh McDermott, and Bevil R. Conway, "Divergence in the Functional Organization of Human and Macaque Auditory Cortex Revealed by fMRI Responses to Harmonic Tones," *Nature Neuroscience* 22 (2019), pp. 1057–60.

16. T. Aran Mooney, Roger T. Hanlon, Jakob Christensen-Dalsgaard, Peter T. Madsen, Darlene R. Ketten, and Paul E. Nachtigall "Sound Detection by the Longfin Squid (*Loligo pealeii*) Studied with Auditory Evoked Potentials: Sensitivity to Low-Frequency Particle Motion and Not Pressure," *Journal of Experimental Biology* 213 (2010), pp. 3748–59.

17. Kevin Healy, Luke McNally, Graeme D. Ruxton, Natalie Cooper, and Andrew L. Jackson, "Metabolic Rate and Body Size Are Linked with Perception of Temporal Information," *Animal Behaviour* 86 (2013), pp. 685–96.

18. Karl Ernst von Baer, *Welche Auffassung der lebenden Natur ist die richtige? Und wie ist diese Auffassung auf die Entomologie anzuwenden?* (Berlin: August Hirschwald, 1862). As with Uexküll's, Baer's conception of life is musical. More pertinently for IMTE, it is fundamentally rhythmic and develops by complex folding patterns of repetition and variation. See Janina Wellmann, *The Form of Becoming: Embryology and the Epistemology of Rhythm, 1760–1830*, trans. Kate Sturge (New York: Zone Books, 2017), pp. 301–320.

19. See Roger S. Payne and Scott McVay's groundbreaking study, "Songs of Humpback

Whales," *Science* 173 (1971), pp. 585–97. Roger Payne also produced the album *Songs of the Humpback Whale.*

20. See Clare Owen, Luke Rendell, Rochelle Constantine, Michael J. Noad, Jenny Allen, Olive Andrews, Claire Garrigue, M. Michael Poole, David Donelly, Nan Hauser, and Ellen C. Garland, "Migratory Convergence Facilitates Cultural Transmission of Humpback Whale Song," *Royal Society Open Science* 6.9 (September 4, 2019).

21. On underwater hearing, see Stefan Helmreich, *Sounding the Limits of Life: Essays in the Anthropology of Biology and Beyond* (Princeton, NJ: Princeton University Press, 2016), pp. 136–53.

22. Margret Grebowicz, *Whale Song* (London: Bloomsbury, 2017), p. 6.

23. See also D. Graham Burnett, *The Sounding of the Whale* (Chicago: University of Chicago Press, 2012), pp. 622–42.

24. SETI started its life as an interdisciplinary group of highly decorated scientists trying to crack the code of nonhuman communication. It is now a nonprofit institute based in Mountain View, California.

25. See John Durham Peters, *The Marvelous Clouds: Toward a Philosophy of Elemental Media* (Chicago: University of Chicago Press, 2015), p. 57.

26. Ibid., p. 66.

27. On the music and the conditions for biosemiotics, see Gary Tomlinson, "Sign, Affect, and Musicking before the Human," *boundary 2* 43.1 (2016), pp. 143–72.

28. As the American neurologist John Cunningham Lilly noted during the human-dolphin cohabitation experiment: "No matter how long it takes, no matter how much work, *this dolphin is going to learn to speak English!*" See John Lilly, *The Mind of the Dolphin: A Nonhuman Intelligence* (New York: Avon, 1967), p. 198. For a critical account of this project and other questionable research by Lilly in relation to the alien other, see Gavin Steingo, *Splendid Universe: Sound, Music, and Interspecies Communication* (Chicago: University of Chicago Press, forthcoming).

29. In his celebrated TV series *Cosmos*, Sagan highlights how our understanding of the world has gradually shifted away from predominantly geocentric and anthropocentric models. Sagan's history proceeds as a series of revolutions that have seemingly shaken our collective and individual sense of importance. The Copernican Revolution was the first Earth-shattering event, and the discovery of exoplanetary civilizations may be the next one coming.

30. Carl Sagan, *Pale Blue Dot: A Vision of the Human Future in Space* (New York: Ballantine, 1997), p. 77.

CHAPTER SIX: DEFINITION

1. The notion that music makes time and is time parallels Deleuze and Guattari's concept of the refrain (*ritournelle*): "The refrain fabricates time (*du temps*)," and is "the a priori form of time." Gilles Deleuze and Félix Guattari, *A Thousand Plateaus: Capitalism and Schizophrenia*, trans. Brian Massumi (Minneapolis: University of Minnesota Press, 1987), p. 349. See also the Introduction, note 17.

2. On the music of bamboo forests, see Timothy Morton's poetic invocation of interobjectivity in *Hyperobjects: Ecology and Philosophy after the End of the World* (Minneapolis: University of Minnesota Press, 2013), p. 81.

3. For details, see Chapters 3 and 5.

4. Aurelius Augustinus, *De musica*, ed. Mark Jacobson (Berlin: De Gruyter, 2017), 1.2.2.

5. The idea of the musical event as a predicate is indebted to Leibniz. For a clear exposition on the Leibnizian event, see Gilles Deleuze, *The Fold: Leibniz and the Baroque*, trans. Tom Conley (Minneapolis: University of Minnesota Press, 1993), pp. 52–54 and 76–82.

6. Gottfried Leibniz, Letter to Christian Goldbach, April 17, 1712. "Music is a hidden arithmetic exercise of the soul, which does not know that it is counting." Original in Gottfried Wilhelm Leibniz, *Opera omnia*, ed. Ludovic Dutens, 6 vols. (1768; Hildesheim; G. Olms, 1989), vol. 3, *Opera mathematica*, pp. 437–38. Transmission is complicated by the fact that the letter states *animae* (of the soul), whereas the printed version, which was widely read, is changed to *animi* (of the spirit). See Ulrich Leisinger, *Leibniz-Reflexe in der deutschen Musiktheorie des 18. Jahrhunderts* (Würzburg: Königshausen und Neumann, 1994), p. 43.

7. Metaphors, as Mark Johnson and George Lakoff point out, are embodied experiences for humans. See George Lakoff and Mark Johnson, *Metaphors We Live By* (Chicago: University of Chicago Press, 1980); Lakoff and Johnson, *Philosophy in the Flesh: The Embodied Mind and Its Challenge to Western Thought* (New York: Basic Books, 1999); and Mark Johnson, *The Body in the Mind: The Bodily Basis of Meaning, Imagination, and Reason* (Chicago: University of Chicago Press, 1987). Johnson and Lakoff's work has been applied to music, particularly to the notions of movement and temporality, by Steve Larson in *Music Forces: Motion, Metaphor, and Meaning in Music* (Bloomington: Indiana University Press, 2012), and Arnie Cox in *Music and Embodied Cognition: Listening, Moving, Feeling, and Thinking* (Bloomington: Indiana University Press, 2016).

8. Graham Harman, *Object-Oriented Ontology: A New Theory of Everything* (London: Penguin, 2018), pp. 61–102. The claims here are indebted to object-oriented ontology.

9. On a similar view of music as both thing and process, see Jonathan Sterne, *MP3: The Meaning of a Format* (Durham, NC: Duke University Press, 2012).

10. Musical semiosis need not be "meaningful" in a symbolic sense. As numerous music semioticians have pointed out, of the three related types of signs defined by Charles Sanders Peirce—icon, index, and symbol—music is primarily indexical. It points to relations. For humans (and by extension, Sagan's "human+" aliens), music's indexicality is caught within a symbolic web of meaning, since human culture is largely distinguished by the symbolic. But as Gary Tomlinson notes, for nonhuman animals (at least on Earth), the index is the sum total of their semiosphere; frequency information creates an indexical network of meaning. Georgina Born, taking her cue from Alfred North Whitehead, suggests that prehension—perception, but not cognition—may be sufficient to trigger a network of becoming between subject and object at any given level; an entity that prehends another object results in a co-creative act that actualizes the potential immanent in the object. Music's external network is a wide spectrum of mediations. See Gary Tomlinson, "Sign, Affect, and Musicking before the Human," *boundary 2* 43.1 (2016), pp. 145–50; and Georgina Born, "On Nonhuman Sounds—Sounds as Relations," in James A. Steintrager and Rey Chow, eds., *Sound Objects* (Durham, NC: Duke University Press, 2019), pp. 185–209.

11. Kant's definition of the beautiful in the *Critique of Judgment*, "Zweckmässigkeit ohne Zweck," is most memorably translated in English by J. H. Bernard in the *Critique of Judgment* as "purposiveness without purpose."

12. The use of "dark" and "bright objects" is somewhat loosely inspired by Levi R. Bryant, *Onto-Cartography: An Ontology of Machines and Media* (Edinburgh: Edinburgh University Press, 2014).

13. See Simon Conway Morris, *Life's Solution: Inevitable Humans in a Lonely Universe* (Cambridge: Cambridge University Press, 2003).

14. On unalienable possessions, see Anna Morcom, "Music, Exchange, and the Production of Value: A Case Study of Hindustani Music," in Anna Morcom and Timothy Taylor, eds., *The Oxford Handbook of Economic Ethnomusicology*, https://doi.org/10.1093/oxfordhb/9780190859633.013.28.

15. Melissa Mueller, "Helen's Hands: Weaving for *Kleos* in the *Odyssey*," *Helios* 37.1 (2010), pp. 1–20.

16. Augustinus, *De musica*, 6.11.29.

CHAPTER SEVEN: REPEAT

The epigraph is from Gilles Deleuze, *The Fold: Leibniz and the Baroque*, trans. Tom Conley (Minneapolis: University of Minnesota Press, 1993), p. 137.

1. See Graham Harman, *Object-Oriented Ontology: A New Theory of Everything* (London: Penguin, 2018), pp. 21–24.

2. Douglas Adams, *The Hitchhiker's Guide to the Galaxy* (New York: Del Rey Books, 2005), p. 34.

3. Carl Sagan, Francis D. Drake, Ann Druyan, Timothy Ferris, Jon Lomberg, and Linda Salzman Sagan, *Murmurs of Earth: The Voyager Interstellar Record* (New York: Random House, 1978), p. 167.

4. From Jimmy Carter's typed statement onboard Voyager.

5. Timothy Ferris's message inscribed in the Golden Record's takeout groove.

6. Fearing the possibility of alien invasions, Astronomer Royal of England and Nobel laureate Martin Ryle even petitioned the Executive Committee of the International Astronomical Union to approve a resolution condemning such use of cosmic maps on Pioneer and Voyager. See Sagan et al., *Murmurs of Earth*, p. 66.

7. Cixin Liu's trilogy *The Three-Body Problem* (of which the middle volume is entitled *Dark Forest*) spins out this violent universe into a tale of epic proportions. The trilogy ends with a message in a bottle from an advanced humanity that makes Voyager's time capsule seem Paleolithic.

8. Violent music exists, of course. And—who knows?—the rhythmic belligerence of the *Rite of Spring* might cause NASA's peace mission to backfire and provoke an alien attack from the direction of our nearest pulsar. Yet as music, Stravinsky's sacrificial dance belies the violence of its death drive: its frequencies encode another message. As the composer himself scribbled in the sketches for the ballet: "Music exists if there is rhythm, as life exists if there is a pulse." And where there is a pulse blinking in the universe, space-time is woven into a musical order. It is as if music cannot help but be a harbinger of peace in all its polyphonic vitality. NASA's gift of music aligns with its mission "to come in peace." The same could be said of Mozart's revenge aria on the Golden Record, where the violent verbal intent of the Queen of the Night is superseded by the sheer virtuosic brilliance of the music. See Igor Stravinsky, *The Rite of Spring: Sketches 1911–13* (London: Boosey and Hawkes, 1969), p. 36, and Robert Craft, "'The Rite of Spring': Genesis of a Masterpiece," in ibid., p. xxxiii.

9. See Friedrich Nietzsche, *The Birth of Tragedy out of the Spirit of Music*, trans. Ian Johnston (Arlington, VA: Richer Resources Publications, 2009), §21, p. 109. To use NASA's

terms, Nietzsche's aesthetic vision combines the Apollo program with the space shuttle. Taking his cue from Wagner, Nietzsche primarily associated music with Dionysian harmony, its overwhelming forces representing Schopenhauer's cosmic will; rhythm belonged to the ordered and individualized realm of Apollo.

10. See David Grinspoon, *Earth in Human Hands: Shaping Our Planet's Future* (New York: Grand Central Publishing, 2016), pp. 305–51. Also see Dipesh Chakrabarty, "The Human Condition in the Anthropocene," The Tanner Lectures in Human Values, New Haven, Yale University, delivered February 18–19, 2015, available at https://tannerlectures.utah.edu/Chakrabarty%20manuscript.pdf.

11. Wagner's music was deliberately excluded from the Golden Record because of its association with Hitler, as were recordings by Elisabeth Schwarzkopf and Herbert von Karajan. However, in a twist of cosmic irony, United Nations Secretary General Kurt Waldheim, who delivers the first message on the record (see note 12 below), was later discovered to have been a wartime intelligence officer in the Nazi regime. See Jonathan Scott, *The Vinyl Frontier: The Story of the Voyager Golden Record* (London: Bloomsbury, 2019), pp. 237–38.

12. From the message on Voyager from Secretary General of the United Nations Kurt Waldheim.

13. In the *Odyssey* (books 5 and 10), both Circe and Calypso sing while they weave. On the loom as lyre, see Jane M. Snyder, "The Web of Song: Weaving Imagery in Homer and the Lyric Poets," *Classical Journal* 76.3 (1981), p. 194. On singing and the computation of complex patterns on fabric, see Anthony Tuck, "Singing the Rug: Patterned Textiles and the Origins of Indo-European Metrical Poetry," *American Journal of Archaeology* 110.4 (2006), pp. 539–50. Also see the insert on Penelope in Chapter 2.

14. This is the function of Deleuze and Guattari's *ritournelle* (refrain) in *A Thousand Plateaus*. Plateau 11, "Of the Refrain," begins with the example of a child singing to himself in the dark to assuage his fear and bring order into chaos. Gilles Deleuze and Félix Guattari, *A Thousand Plateaus: Capitalism and Schizophrenia*, trans. Brian Massumi (Minneapolis: University of Minnesota Press, 1987), p. 311. See also the final endnote of the Introduction in this volume.

15. In the *Odyssey*, book 24, one of the suitors, Amphimedon, retells the weaving story in Hades, revealing that the funeral shroud that Penelope was eventually forced to finish was ironically destined for her suitors and brought about their destruction. See Steven Lowenstam, "The Shroud of Laertes and Penelope's Guile," *Classical Journal* 95.4 (2000), pp. 333–48.

16. Penelope's weaving is often associated with her intelligence and cunning (*mêtis*); the weave symbolizes a twisting and plotting that is the doing and undoing of those in her web. On weaving and *mêtis* in general, see Ann Bergren, "Language and the Female in Early Greek Thought," *Arethusa* 16.1–2 (1983), pp. 71–75; on Penelope's weaving and *mêtis* in particular, see Barbara Clayton, *A Penelopean Poetics: Reweaving the Feminine in Homer's* Odyssey (Oxford: Lexington Books, 2004).

17. See, for example, Emmanuel Lévinas, "Diachrony and Representation," in *Entre Nous: Essays on Thinking-of-the-Other*, trans. Michael B. Smith and Barbara Harshav (New York: Columbia University Press, 1993), pp. 159–77.

18. Raimon Panikkar, *The Rhythm of Being: The Unbroken Trinity* (Maryknoll, NY: Orbis Books, 2010), chapter 1, section 3C, loc. 1469.

19. Carl Sagan, quoted at https://voyager.jpl.nasa.gov/golden-record/whats-on-the-record.

Readings

Adams, Douglas. *The Hitchhiker's Guide to the Galaxy*. New York: Del Rey Books, 2005.

Adorno, Theodor W. *Aesthetic Theory*. Translated by Robert Hullot-Kentor. Minneapolis: University of Minnesota Press, 1997.

———. "Beethoven's Late Style." In Rolf Tiedemann, ed., *Beethoven: The Philosophy of Music*. Translated by Edmund Jephcott. Cambridge: Polity, 1998, pp. 123–26.

———. "Subject and Object." In Andrew Arato and Eike Gebhardt, eds., *The Essential Frankfurt School Reader*. New York: Continuum, 1982, pp. 497–510.

———, and Max Horkheimer. *Dialectic of Enlightenment*. Translated by John Cumming. London: Verso, 1979.

Agawu, Kofi. "Tonality as a Colonizing Force in Africa." In Ronald Radano, ed., *Audible Empire: Music, Global Politics, Critique*. Durham, NC: Duke University Press, 2018, pp. 334–55.

Arnim, Bettina von, and Karoline von Arnim. *Die Günderode*. Frankfurt am Main: Insel, 1982.

Arthur, Richard. T. W. *Leibniz*. Cambridge: Polity, 2014.

Augustinus, Aurelius. *De musica*. Edited by Mark Jacobson. Berlin: De Gruyter, 2017.

Baer, Karl E. von. *Welche Auffassung der lebenden Natur ist die richtige? Und wie ist diese Auffassung auf die Entomologie anzuwenden?* Berlin: August Hirschwald, 1862.

Bai, Junxiao. "Numbers: Harmonic Ratios and Beauty in Augustinian Musical Cosmology." *Cosmos and History* 13.3 (2017), pp. 192–217.

Barry, Ron. "How to Decode the Images on the Voyager Golden Record," *Boing Boing*, September 5, 2017, https://boingboing.net/2017/09/05/how-to-decode-the-images-on-th.html.

Barthes, Roland. *Camera Lucida: Reflections on Photography*. Translated by Richard Howard. New York: Hill and Wang, 2010.

Bennett, Jane. *Vibrant Matter: A Political Ecology of Things*. Durham, NC: Duke University Press, 2010.

Bergren, Ann. "Language and the Female in Early Greek Thought." *Arethusa* 16.1–2 (1983), pp. 69–95.

Boethius, Anicius Manlius Severinus. *Fundamentals of Music*. Edited and translated by Calvin M. Bower and Claude V. Palisca. New Haven, CT: Yale University Press, 1989.

Bogost, Ian. *Alien Phenomenology, or What It's Like to Be a Thing*. Minneapolis: University of Minnesota Press, 2012.

———. *Play Anything: The Pleasure of Limits, the Uses of Boredom, and the Secret of Games* New York: Basic Books, 2016.

Bohlman, Philip, ed. *The Cambridge History of World Music*. Cambridge: Cambridge University Press, 2014.

Bonds, Mark Evan. *Absolute Music: The History of an Idea*. New York: Oxford University Press, 2014.

Born, Georgina. "On Nonhuman Sounds—Sounds as Relations." In James A. Steintrager and Rey Chow, eds., *Sound Objects*. Durham, NC: Duke University Press, 2019, pp. 185–209.

Boym, Svetlana. *The Future of Nostalgia*. New York: Basic Books, 2004.

Brennan, Timothy. "World Music Does Not Exist." *Discourse* 23.1 (2001), pp. 44–62.

Brown, Bill. "Thing Theory." *Critical Inquiry* 28.1 (2001), pp. 1–22.

Bryant, Levi R. *Onto-Cartography: An Ontology of Machines and Media*. Edinburgh: Edinburgh University Press, 2014.

Burnett, D. Graham. *The Sounding of the Whale*. Chicago: University of Chicago Press, 2012.

Cagniard de la Tour, Charles. "Sur la sirène, nouvelle machine d'acoustique pour mesurer les vibrations de l'air qui constituent le son." *Annales de la physique et chimie* 12 (1819), pp. 167–71.

Chakrabarty, Dipesh. "The Human Condition in the Anthropocene." The Tanner Lectures in Human Values, New Haven, Yale University, delivered February 18–19, 2015, https://tannerlectures.utah.edu/Chakrabarty%20manuscript.pdf.

Chua, Daniel K.L. *Absolute Music and the Construction of Meaning*. Cambridge: Cambridge University Press, 1999.

_____. "Beethoven's Other Humanism." *Journal of the American Musicological Society* 62.3 (2009), pp. 571–645.

_____. "In Time with the Late Style." Preface in Leo Ou-fan Lee and Henry Shiu, 【諸神的黃昏】–晚期風格的跨學科對談: *In Time with the Late Style*. Hong Kong: Oxford University Press, 2019.

_____. "Rioting with Stravinsky: A Particular Analysis of the *Rite of Spring*." *Music Analysis* 26.1–2 (2007), pp. 59–109.

Clayton, Barbara. *A Penelopean Poetics: Reweaving the Feminine in Homer's* Odyssey. Oxford: Lexington Books, 2004.

Coccia, Emanuele. *The Life of Plants: A Metaphysics of Mixture*. Cambridge: Polity, 2019.

Conway Morris, Simon. *Life's Solution: Inevitable Humans in a Lonely Universe*. Cambridge: Cambridge University Press, 2003.

Cox, Arnie. *Music and Embodied Cognition: Listening, Moving, Feeling, and Thinking*. Bloomington: Indiana University Press, 2016.

Craft, Robert. "'The Rite of Spring': Genesis of a Masterpiece." In Igor Stravinsky, *The Rite of Spring: Sketches 1911–13*. London: Boosey and Hawkes, 1969.

Cross, Ian. "Is Music the Most Important Thing We Ever Did?: Music, Development, and Evolution." In Suk Won Yi, ed., *Music, Mind, and Science*. Seoul: Seoul National University Press, 1999, pp. 10–39.

Dahlhaus, Carl. *The Idea of Absolute Music*. Translated by Roger Lustig. Chicago: University of Chicago Press, 1989.

Delanda, Manuel. *Assemblage Theory*. Edinburgh: Edinburgh University Press, 2016.

Delanda, Manuel, and Graham Harman. *The Rise of Realism*. Cambridge: Polity, 2017.

Deleuze, Gilles. *Difference and Repetition*. Translated by Paul Patton. New York: Columbia University Press, 1994.

_____. *The Fold: Leibniz and the Baroque*. Translated by Tom Conley. Minneapolis: University of Minnesota Press, 1993.

_____, and Félix Guattari. *A Thousand Plateaus: Capitalism and Schizophrenia*. Translated by Brian Massumi. Minneapolis: University of Minnesota Press, 1987.

Dell'Antonio, Andrew, ed. *Beyond Structural Listening: Postmodern Modes of Hearing*. Berkeley: University of California Press, 2004.

Derrida, Jacques. *Of Grammatology*. Corrected edition. Translated by Gayatri Chakravorty Spivak. Baltimore: Johns Hopkins University Press, 1998.

Deutsch, Diana, Trevor Henthorn, and Rachael Lapidis. "Illusory Transformation from Speech to Song." *Journal of the Acoustical Society of America* 129.4 (2011), pp. 2245–52.

Diderot, Denis. *Diderot's Selected Writings.* Edited by Lester G. Crocker. Translated by Derek Coltman. New York: Macmillan, 1966.

Dolan, Emily. "Toward a Musicology of Interfaces." *Keyboard Perspectives* 5 (2012), pp. 1–13.

Druyan, Ann. "Dear Voyagers: How Your Billion-Year Journey Carries True Love." *National Geographic*, July 10, 2019.

Eikelboom, Lexi. *Rhythm: A Theological Category.* Oxford: Oxford University Press, 2018.

Erlmann, Veit. *Reason and Resonance: A History of Modern Aurality.* New York: Zone, 2010.

Ernst, Wolfgang. *Digital Media and the Archive.* Translated by Jussi Parikka. Minneapolis: University of Minnesota Press, 2012.

———. *Im Medium erklingt die Zeit: Technologische Tempor(e)alitäten.* Berlin: Kulturverlag Kadmos, 2015. English revised translation as *Sonic Time Machines, Explicit Sound, Sirenic Voices, and Implicit Sonicity.* Amsterdam: Amsterdam University Press, 2016.

Ferris, Timothy. "How the Golden Record Was Made." *New Yorker*, August 20, 2017.

Flusser, Vilém, and Louis Bec. *Vampyroteuthis Infernalis: A Treatise, with a Report by the Institut Scientifique de Recherche Paranaturaliste.* Translated by Valentine Pakis. Minneapolis: University of Minnesota Press, 2012.

Forte, Allen. *The Harmonic Organization of "The Rite of Spring."* New Haven, CT: Yale University Press, 1978.

———. *The Structure of Atonal Music.* New Haven, CT: Yale University Press, 1973.

Gallope, Michael. *Deep Refrains: Music, Philosophy, and the Ineffable.* Chicago: University of Chicago Press, 2017.

Galloway, Alexander R. *The Interface Effect.* Cambridge: Polity Press, 2012.

Garcia, Luis-Manuel. "On and On: Repetition as Process and Pleasure in Electronic Dance Music." *Music Theory Online* 11.4 (2005), https://mtosmt.org/issues/mto.05.11.4/mto.05.11.4.garcia.html.

Garcia, Tristan. *Form and Object: A Treatise on Things.* Translated by Mark Allan Ohm and Jon Cogburn. Edinburgh: Edinburgh University Press, 2014.

Gell, Alfred. *Art and Agency: An Anthropological Theory.* Oxford: Oxford University Press, 1998.

Godfrey-Smith, Peter. *Other Minds: The Octopus, the Sea, and the Deep Origins of Consciousness.* London: Farrar, Straus and Giroux, 2016.

Gorman, Alice. "Beyond the Morning Star: The Real Tale of the Voyager's Aboriginal Music." *The Conversation*, October 2, 2013, https://theconversation.com/beyond-the-morning-star-the-real-tale-of-the-voyagers-aboriginal-music-18288.

Gould, Stephen Jay. *Wonderful Life: The Burgess Shale and the Nature of History*. New York: W. W. Norton, 1990.

Grebowicz, Margaret. *Whale Song*. New York: Bloomsbury, 2017.

Griffin, Donald R. *Listening in the Dark: The Acoustic Orientation of Bats and Men*. New York: Dover, 1974.

Grinspoon, David. *Earth in Human Hands: Shaping Our Planet's Future*. New York: Grand Central Publishing, 2016.

Grusin, Richard, ed. *The Nonhuman Turn*. Minneapolis: University of Minnesota Press, 2015.

Gulik, R. H. van. "The Lore of the Chinese Lute: An Essay in Ch'in Ideology." *Monumenta Nipponica* 1.2 (1938), pp. 386–438.

Haffke, Maren. *Archäologie der Tastatur: Musikalische Medien nach Friedrich Kittler und Wolfgang Scherer*. Munich: Wilhelm Fink, 2019.

Hagen, Wolfgang. "Metaxy: Eine historiosemantische Fußnote zum Medienbegriff." In Stefan Münzler and Alexander Rösler, eds., *Was ist ein Medium?* Frankfurt am Main: Suhrkamp, 2005.

Hansen, Mark B. N. "Symbolizing Time: Kittler and Twenty-First Century Media." In Stephen Sale and Laura Salisbury, eds., *Kittler Now: Current Perspectives in Kittler Studies*. Cambridge: Polity, 2015, pp. 210–37.

Harman, Graham. *Guerrilla Metaphysics: Phenomenology and the Carpentry of Things*. Chicago: Open Court, 2005.

———. *Object-Oriented Ontology: A New Theory of Everything*. London: Penguin, 2018.

———. *Speculative Realism: An Introduction*. Cambridge: Polity Press, 2018.

———. *Tool-Being: Heidegger and the Metaphysics of Objects*. Chicago: Open Court, 2002.

Harrison, Carol. "Getting Carried Away: Why Did Augustine Sing?" *Augustine Studies* 46.1 (2015), pp. 1–22.

Hasty, Christopher F. *Meter as Rhythm*. Oxford: Oxford University Press, 1997.

Healy, Kevin, Luke McNally, Graeme D. Ruxton, Natalie Cooper, and Andrew L. Jackson. "Metabolic Rate and Body Size Are Linked with Perception of Temporal Information." *Animal Behaviour* 86 (2013), pp. 685–96.

Heidegger, Martin. *Being and Time*. Translated by John Macquarrie and Edward Robinson. New York: Harper & Row, 1962.

Heller-Roazen, Daniel. *The Fifth Hammer: Pythagoras and the Disharmony of the World*. New York: Zone Books, 2011.

Helmholtz, Hermann. "On the Physiological Causes of Musical Harmony (1857)." In David Cahan, ed., *Science and Culture*. Chicago: University of Chicago Press, 1995, pp. 46–57.

———. *On the Sensations of Tone*. Translated by Alexander Ellis. New York: Dover, 1954.

Helmreich, Stefan. "Music for Cochlear Implants." In Emily Dolan and Alexander Rehding, eds., *The Oxford Handbook of Timbre*. New York: Oxford University Press, 2020.

———. *Sounding the Limits of Life: Essays in the Anthropology of Biology and Beyond*. Princeton: Princeton University Press, 2016.

Hepokoski, James. *Sibelius: Symphony No. 5*. Cambridge: Cambridge University Press, 1993.

———. "Structure, Implication, and the End of *Suor Angelica*." *Studi pucciniani* 3 (2004), pp. 241–64.

Herder, Johann Gottfried. *Werke*. Vol. 8, *Schriften zu Literatur und Philosophie, 1792–1800*. Edited by Hans Dietrich Irmscher. Frankfurt am Main: Deutscher Klassiker Verlag, 1998.

Hildegard of Bingen. *Scivias*. Translated by Columba Hart and Jane Bishop. New York: Paulist Press, 1990.

Hoerner, Sebastian von. *Sind wir allein?: SETI und das Leben im All*. Munich: C. H. Beck, 2003.

———. "Universal Music?" *Psychology of Music* 2.2 (1974), pp. 18–28.

Ingold, Tim. *The Life of Lines*. London: Routledge, 2015.

———. *Lines: A Brief History*. London: Routledge, 2016.

———. *The Perception of the Environment: Essays on Livelihood, Dwelling and Skill*. London: Routledge, 2000.

Jensen, Bas. "Tape Cassettes and Former Selves: How Mix Tapes Mediate Memories." In Karin Bijsterveld and José van Dijck, eds., *Sound Souvenirs: Audio Technologies, Memory, and Cultural Practices*. Amsterdam: Amsterdam University Press, 2009, pp. 43–54.

Johnson, Mark. *The Body in the Mind: The Bodily Basis of Meaning, Imagination, and Reason*. Chicago: University of Chicago Press, 1987.

Kaufmann, Walter. *Musical Notations of the Orient*. Indianapolis: Indiana University Press, 1967.

Kierkegaard, Søren. *Repetition and Philosophical Crumbs*. Translated by M. G. Piety. Oxford: Oxford University Press, 2009.

Kittler, Friedrich. *Gramophone, Film, Typewriter*. Translated by Geoffrey Winthrop-Young and Michael Wutz. Stanford, CA: Stanford University Press, 1999.

Kivy, Peter. *The Fine Art of Repetition: Essays in the Philosophy of Music*. Cambridge: Cambridge University Press, 1993.

Korsyn, Kevin. *Decentering Music: A Critique of Contemporary Musical Research*. New York: Oxford University Press, 2003.

Kramer, Lawrence. *The Hum of the World: A Philosophy of Listening*. Oakland: University of California Press, 2018.

Krämer, Sybille. "The Cultural Techniques of Time Axis Manipulation: On Friedrich Kittler's Conception of Media." *Theory, Culture & Society* 23.7–8 (2006), pp. 93–109.

Kruger, Kathryn Sullivan. *Weaving the Word: The Metaphorics of Weaving and Female Textual Production*. Selinsgrove, PA: Susquehanna University Press, 2001.

Lakoff, George, and Mark Johnson. *Metaphors We Live By*. Chicago: University of Chicago Press, 1980.

———. *Philosophy in the Flesh: The Embodied Mind and Its Challenge to Western Thought*. New York: Basic Books, 1999.

Larson, Steve. *Music Forces: Motion, Metaphor, and Meaning in Music*. Bloomington: Indiana University Press, 2012.

Lefebvre, Henri. *The Production of Space*. Translated by Donald Nicholson-Smith. Oxford: Blackwell, 1991.

Leibniz, Gottfried Wilhelm. *Opera omnia*. Vol. 3, *Opera mathematica*. Edited by Ludovic Dutens. 6 vols. 1768; Hildesheim: G. Olms, 1989.

———. *Philosophical Essays*. Translated by Roger Ariew and Daniel Garber. Indianapolis: Hackett, 1989.

———, *Die Philosophischen Schriften*. Edited by C. I. Gerhardt. 1875–90; Hildesheim: Olms, 1960–61.

———. *Protogaea*. Translated by Claudine Cohen and Andre Wakefield. Chicago: University of Chicago Press, 2008.

Leisinger, Ulrich. *Leibniz-Reflexe in der deutschen Musiktheorie des 18. Jahrhunderts*. Würzburg: Königshausen und Neumann, 1994.

Lévinas, Emmanuel. *Alterity and Transcendence*. Translated by Michael B. Smith. London: Athlone, 1999.

———. *Entre Nous: On-Thinking-of-the-Other*. Translated by Michael B. Smith and Barbara Harshav. New York: Columbia University Press, 1993.

———. *Totality and Infinity*. Translated by Alphonso Lingis. Pittsburgh: Duquesne University Press, 1969.

Lilly, John C. *The Mind of the Dolphin: A Nonhuman Intelligence*. New York: Avon, 1967.

Lowenstam, Steven. “The Shroud of Laertes and Penelope’s Guile.” *Classical Journal* 95.4 (2000), pp. 333–48.

Luhmann, Niklas. “The Cognitive Program of Constructivism and the Reality That Remains Unknown.” In William Rasch, ed., *Theories of Distinction: Redescribing the Descriptions of Modernity*. Stanford, CA: Stanford University Press, 2002.

———. “The Improbability of Communication.” *International Social Science Journal* 23.1 (1981), pp. 122–32.

Mailman, Joshua Banks. “Cybernetic Phenomenology of Music, Embodied Speculative Realism, and Aesthetic-Driven Techné for Spontaneous Audio-Visual Expression.” *Perspectives of New Music* 54.1 (2016), pp. 5–95.

Marchetti, Christopher C. “Aristoxenus’ *Elements of Rhythm*: Text, Translation, and Commentary with a Translation and Commentary on *POxy* 2687.” PhD diss., Rutgers, 2009.

Margulis, Elizabeth Hellmuth. *On Repeat: How Music Plays the Mind*. Oxford: Oxford University Press, 2014.

Mitchell, W. T. J. “Ekphrasis and the Other.” In *Picture Theory: Essays on Verbal and Visual Representation*. Chicago: University of Chicago Press, 1994, pp. 151–81.

Mooney, T. Aran, Roger T. Hanlon, Jakob Christensen-Dalsgaard, Peter T. Madsen, Darlene R. Ketten, and Paul E. Nachtigall. “Sound Detection by the Longfin Squid (*Loligo pealeii*) Studied with Auditory Evoked Potentials: Sensitivity to Low-Frequency Particle Motion and Not Pressure.” *Journal of Experimental Biology* 213 (2010), pp. 3748–59.

Morcom, Anna. “Music, Exchange, and the Production of Value: A Case Study of Hindustani Music.” In Anna Morcom and Timothy Taylor, eds., *The Oxford Handbook of Economic Ethnomusicology*, doi:10.1093/oxfordhb/9780190859633.001.0001.

Morena, Anthony Michael. *The Voyager Record: A Transmission*. Brookline, MA: Rose Metal Press, 2016.

Morton, Timothy. *Hyperobjects: Ecology and Philosophy after the End of the World*. Minneapolis: University of Minnesota Press, 2013.

Moseley, Roger. *Keys to Play: Music as a Ludic Medium from Apollo to Nintendo*. Oakland: University of California Press, 2016.

Mueller, Melissa. "Helen's Hands: Weaving for *Kleos* in the *Odyssey*." *Helios* 37.1 (2010), pp. 1–20.

Nagel, Thomas. "What Is It Like to Be a Bat?" *Philosophical Review* 83 (1974), pp. 435–50.

Nelson, Stephanie, and Larry Polansky. "The Music of the Voyager Interstellar Record." *Journal of Applied Communication Research* 21.4 (1993), pp. 358–76.

Nicomachus of Gerasa, *The Manual of Harmonics of Nicomachus the Pythagorean*. Edited and translated by Flora R. Levin. Grand Rapids, MI: Phanes Press, 1994.

Nietzsche, Friedrich. *The Birth of Tragedy out of the Spirit of Music*. Translated by Ian Johnston. Arlington, VA: Richer Resources Publications, 2009.

———. *The Gay Science*. Translated by Walter Kaufmann. New York: Vintage, 1974.

Norman-Haignere, Sam V., Nancy Kanwisher, Josh McDermott, and Bevil R. Conway. "Divergence in the Functional Organization of Human and Macaque Auditory Cortex Revealed by fMRI Responses to Harmonic Tones." *Nature Neuroscience* 22 (2019), pp. 1057–60.

Novak, David, and Matt Sakakeeny, eds. *Keywords in Sound*. Durham, NC: Duke University Press, 2015.

Oberhaus, Daniel. *Extraterrestrial Languages*. Cambridge, MA: MIT Press, 2019.

Ochoa, Ana Maria. "Exchange: The Geopolitics of Ethnographic Recordings, Music, and Sound." In Naomi Waltham-Smith and Alexander Rehding, eds., *A Cultural History of Western Music*. Vol. 5, *The Industrial Age*. London: Bloomsbury, forthcoming.

O'Flaherty, Wendy Doniger, trans. *The Rig Veda*. London: Penguin Classics, 1981.

Owen, Clare, Luke Rendell, Rochelle Constantine, Michael J. Noad, Jenny Allen, Olive Andrews, Claire Garrigue, M. Michael Poole, David Donelly, Nan Hauser, and Ellen C. Garland. "Migratory Convergence Facilitates Cultural Transmission of Humpback Whale Song." *Royal Society Open Science* 6.9 (September 4, 2019), doi.org/10.1098/rsos.190337.

Panikkar, Raimon. *The Rhythm of Being: The Unbroken Trinity*. Maryknoll, NY: Orbis Books, 2010.

Parikka, Jussi. *What is Media Archaeology?* Cambridge: Polity, 2012.

———, and Erkki Huhtamo, eds., *Media Archaeology*. Berkeley: University of California Press, 2011.

Patel, Aniruddh D. "Evolutionary Music Cognition: Cross-Species Studies." In p. Jason Rentfrow and Daniel Levitin, eds., *Foundations in Music Psychology: Theory and Research*. Cambridge, MA: MIT Press, 2019, pp. 459–501.

Payne, Roger S., and Scott McVay. "Songs of Humpback Whales." *Science* 173 (1971), pp. 585–97.

Pepys, Samuel. *Diary of Samuel Pepys: A New and Complete Transcription*. Edited by Robert Latham and William Matthews. London: Bell, 1970.

Peters, Ben, ed. *Digital Keywords: A Vocabulary of Information Society and Culture*. Princeton, NJ: Princeton University Press, 2016.

Peters, John Durham. *The Marvelous Clouds: Toward a Philosophy of Elemental Media*. Chicago: University of Chicago Press, 2015.

Pickstock, Catherine. *Repetition and Identity*. Oxford: Oxford University Press, 2013.

Pinker, Steven. *How the Mind Works*. New York: Norton, 1997.

Price, Evander. "Future Monumentality." PhD diss., Harvard University, 2019.

Rahn, John. "Repetition," *Contemporary Music Review* 7.2 (1993), pp. 49–57.

Rehding, Alexander. *Beethoven's Symphony No. 9*. New York: Oxford University Press, 2017.

———. "Consonance and Dissonance." In Alexander Rehding and Steven Rings, eds., *The Oxford Handbook of Critical Concepts in Music Theory*. New York: Oxford University Press, 2019, pp. 437–66.

———. "The Dream of a Universal Notation: An Appreciation." *Analytical Approaches to World Music* 8.2 (2020), pp. 314–23.

———. "Of Sirens Old and New." In Jason Stanyek and Sumanth Gopinath, eds., *The Oxford Handbook of Mobile Music Studies, Volume 2*. New York: Oxford University Press, 2014, pp. 77–106.

———. "Opelt's Siren and the Technologies of Musical Hearing." In Viktoria Tkaczyk, Mara Mills, and Alexandra Hui, eds., *Testing Hearing: The Making of Modern Aurality*. New York: Oxford University Press, 2020, pp. 131–57.

Sagan, Carl. *The Cosmic Connection: An Extraterrestrial Perspective*. Garden City, NY: Anchor Books, 1980.

———. *Pale Blue Dot: A Vision of the Human Future in Space*. New York: Ballantine, 1997.

———, Francis D. Drake, Ann Druyan, Timothy Ferris, Jon Lomberg, and Linda Salzman

Sagan. *Murmurs of Earth: The Voyager Interstellar Record.* New York: Random House, 1978.

Saussy, Haun. "Exquisite Cadavers Stitched from Fresh Nightmares: Of Memes, Hives, and Selfish Genes." In Haun Saussy, ed., *Comparative Literature in an Age of Globalization.* Baltimore: Johns Hopkins University Press, 2006, pp. 3–42.

Schaeffer, Pierre. *In Search of a Concrete Music.* Translated by Christine North and John Dack. Berkeley: University of California Press, 2012.

Schafer, R. Murray. *The Soundscape: Our Sonic Environment and the Tuning of the Word.* Rochester, VT: Destiny Books, 1999.

Schenker, Heinrich. *Beethovens Fünfte Sinfonie: Eine Darstellung des musikalischen Inhaltes nach der Handschrift unter Berücksichtigung des Vortrages und der Literatur.* Vienna: Universal Edition, 1924.

Schneider, Birgit. *Textiles Prozessieren.* Zurich: Diaphanes, 2007.

Scott, Jonathan. *The Vinyl Frontier: The Story of the Voyager Golden Record.* London: Bloomsbury, 2019.

Shklovskii, Iosif S., and Carl Sagan. *Intelligent Life in the Universe.* Reprint. Boca Raton, FL: Emerson-Adams, 1998.

Siegert, Bernhard. *Passage des Digitalen: Zeichenpraktiken der neuzeitlichen Wissenschaft 1500–1900.* Berlin: Brinkmann und Bose, 2003.

Sloterdijk, Peter. *Kopernikanische Mobilmachung und Ptolemäische Abrüstung: Ästhetischer Versuch.* Frankfurt am Main: Suhrkamp, 1987.

Snyder, Jane M. "The Web of Song: Weaving Imagery in Homer and the Lyric Poets." *Classical Journal* 76.3 (1981), pp. 193–96.

Steingo, Gavin. *Splendid Universe: Sound, Music, and Interspecies Communication.* Chicago: University of Chicago Press, forthcoming.

Steintrager, James A., and Rey Chow, eds. *Sound Objects.* Durham, NC: Duke University Press, 2019.

Sterne, Jonathan. *The Audible Past: Cultural Origins of Sound Reproduction.* Durham, NC: Duke University Press, 2003.

———. *MP3: The Meaning of a Format.* Durham, NC: Duke University Press, 2012.

Stewart, Susan. *On Longing: Narratives of the Miniature, the Gigantic, the Souvenir, and the Collection.* Durham, NC: Duke University Press, 1992.

Stobart, Henry. "Devils, Dreams, and Desire: Siren Traditions and Musical Creation in the Central Southern Andes." In Linda Phyllis Austern and Inna Naroditskaya, eds., *Music of the Sirens*. Bloomington: Indiana University Press, 2006.

Stockhausen, Karlheinz. *Stockhausen on Music: Lectures and Interviews*, ed. Robin Maconie. London: Marion Boyars, 1989.

Stravinsky, Igor. *The Rite of Spring: Sketches 1911–13*. London: Boosey and Hawkes, 1969.

Szendy, Peter. *Kant in the Land of Extraterrestrials: Cosmopolitical Philosofictions*. Translated by Will Bishop. New York: Fordham University Press, 2013.

Taruskin, Richard. "A Myth of the Twentieth Century: *The Rite of Spring*, the Tradition of the New, and 'the Music Itself.'" *Modernism/Modernity* 2.1 (2002), pp. 1–26.

Tomlinson, Gary. *Culture and the Course of Human Evolution*. Chicago: University of Chicago Press, 2018.

———. *One Million Years of Music: The Emergence of Human Modernity*. New York: Zone Books, 2015.

———. "Sign, Affect, and Musicking before the Human." *boundary 2* 43.1 (2016), pp. 143–72.

Trumpener, Katie. "World Music World Literature: A Geopolitical View." In Haun Saussy, ed., *Comparative Literature in an Age of Globalization*. Baltimore: Johns Hopkins University Press, 2006, pp. 183–202.

Tuck, Anthony. "Singing the Rug: Patterned Textiles and the Origins of Indo-European Metrical Poetry." *American Journal of Archaeology* 110.4 (2006), pp. 539–50.

Turkle, Sherry, ed. *Evocative Objects: Things We Think With*. Cambridge, MA: MIT Press, 2007.

Uexküll, Jakob von. *A Foray into the Worlds of Animals and Humans: With a Theory of Meaning*. Translated by Joseph D. O'Neil. Minneapolis: University of Minnesota Press, 2010.

Watkins, Holly. *Musical Vitalities: Ventures in a Biotic Aesthetics of Music*. Chicago: University of Chicago Press, 2018.

Webb, Stephen. *If the Universe Is Teeming with Aliens, Where Is Everybody?* Heidelberg: Springer, 2015.

Wellmann, Janina. *The Form of Becoming: Embryology and the Epistemology of Rhythm, 1760–1830*. Translated by Kate Sturge. New York: Zone Books, 2017.

Werckmeister, Andreas. *Musicalische Paradoxal-Discourse: A Well-Tempered Universe*. Translated by Dietrich Bartel. 1707; Lanham, MD: Lexington Book, 2018.

Zuckerkandl, Victor. *Sound and Symbol: Music and the External World*. Translated by Willard R. Trask. Princeton, NJ: Princeton University Press, 1956.

Index

Zone Books series design by Bruce Mau
Typesetting by Meighan Gale
Image placement and production by Julie Fry
Printed and bound by Maple Press